Calm Frenzy
One Man's Vietnam War

Loring M. Bailey Jr.

With Gratitude

Loring M. Bailey Sr.
Dorothy Bailey
Rik Carlson
Yvette Dion Toohey
David Toohey
Susan Taylor Daugherty
Soren Sorensen
Elizabeth & Peter Sorensen
Chip Lamb
Stephen Minot
Barbara Carlson
David Carlson
David Smith
The Pomfret School
Trinity College
Michael Thurston
Michael Olson
Kim Butterfield
Connor Finnerty

Calm Frenzy: One Man's Vietnam War

By Loring M. Bailey, Jr.

www.calmfrenzy.com

All rights reserved. No part of this book may be used or reproduced in any manner whatsoever without written permission except in the case of brief quotations embodied in critical articles and reviews.

ISBN: 0-9721649-2-8

Published by monkeyswithwings/publishing

Burlington, Vermont

www.monkeyswithwings.com

Maris

Ring

Foreword

Receiving a hand-written letter is becoming a rarity in today's world of email, text messages and social media, but the experience remains one that is unique because it is so intimate. When I open a letter that is addressed to me, I hold evidence of the interior life of another person. When I compose and send a letter, I give something of myself that is much greater than the meaning of the words I have written.

Loring M. Bailey, Jr. wrote prolifically throughout his experience of training and war. With diplomas from Pomfret School and Trinity College in hand, Ring found a passion in the written word. His letters represent the emergence of a new voice. A voice heard only briefly and from the most horrific of circumstances.

Through the kindness and trust of Loring's family and friends, I first read these letters in the spring of 2011 and began to adapt them for the stage. For the next two years, I tried to hear Loring's voice and allow it to lead me toward a theatrical life. The play, 70lbs of Books, was performed by students, faculty and staff of Pomfret School in May of 2013. Quite suddenly, everyone at the school knew Loring Bailey through the power of his own words.

Though Calm Frenzy is the extraordinary legacy of one young soldier, his work honors the countless men and women who have stepped into harm's way to keep us safe. This book is dedicated to all veterans, past, present and future. They all have stories to tell.

Are we listening?

Chip Lamb
Pomfret School
July 2015

Introduction

Loring M. Bailey Jr. graduated from Trinity College, Hartford Connecticut, in 1967. The recipient of writing awards in fiction, a graphic artist, and an ardent automobile enthusiast, Ring continued with graduate courses in English, and was a technical writer for the Electric Boat Division of General Dynamics Corporation. His love for automobiles was seen through pen and ink sketches, water colors, short stories on racing, a collection of plastic and metal die cast replicas, and an extensive library of volumes on development, history and design. With a perennial thirst for good writing, Ring was a vigorous reader, and a strong admirer of Ernest Hemingway.

In the spring of 1968, under the threat of the draft, Ring enlisted in the Army with the option for Officers Candidate School. From his induction in June 1968 and Basic training in Fort Knox Kentucky, began a series of letters to his family, fiance, and friends that became his only dependable means of communication. With an Infantry assignment, basic training led to nine weeks of Advanced Infantry Training in the swamps of Fort Polk Louisiana, "Tigerland: The Home of Infantry Training for Vietnam". While at Polk, arrangements were made through the mail for a Christmas wedding, and on December 23, 1968, Ring was married to Maris Carlson.

Ring continued training with the Officers Candidate School at Fort Belvoir Virginia, where a recurrent knee infection kept him in and out of the base hospital. After six months at Belvoir, he was recycled out of the program and placed on a holdover status, where the final ten months of his Army career was mapped. This included a 39-day leave and nine months of combat infantry in Vietnam.

In early October 1969, Ring reached the war zone, was assigned to the Americal Division, and was stationed at Landing Zone Liz, a forward fire base that operated out of Chu Lai, RVN. Writing stations were established immediately, and the letters flowed regularly out of the jungle.

About mid-November, Ring was assigned as the commanding officers' radio telephone operator, and shortly after that, promoted from private first class to specialist fourth. On the fifteenth of March 1970, while on Nui Chap Vung Mountain in the Quang Ngai province, Ring was fatally wounded from the detonation of a hidden anti-personnel device. He was posthumously awarded the Bronze Star Medal and the Purple Heart.

This volume contains edited excerpts of his letters. They were written to his wife, parents and close friends. To his wife and parents he was often over protective, and the horrors of war were veiled and indirect. He wrote of his physical needs and desires, of the equipment he used, and always optimistically about the future. To friends David and Yvette Toohey, he wrote more directly of the overall military actions with histories and fictions that blended reality with the absurd and drew parallels to the home front. To his brother-in-law Rik, his letters shared an interest in good writing and a love for automobiles.

We who assembled this book felt a need to share what Ring wrote to us. The insight of an articulate and energetic soldier shows Vietnam in the direct perspective of one man in a war with a 360 degree front. We present it to you as it was written to us: in a calm frenzy.

Maris Bailey
Mr. & Mrs Loring M. Bailey Sr.
David & Yvette Toohey
Rik Carlson

TRAINING

Every day this life, which can be so sweet and deep, opens up again and I go into it wearing olive green, and I cannot perceive its taste and depth. To see it I have to drug the pain first and then drug my unboosted senses ... one life, one world, one time, they elude me now. I try to remember.

★ ★ ★

...Here is Kentucky where high school basketball, chrome reverse wheels and short haircuts are king. The Army is a southern institution and it's entirely at home in a state like this one. The soldiers mentality can quickly shift from Bear Bryants Football in Alabama to the triple-A basketball runoff, all in a white air-conditioned Pontiac Catalina and a whitewall haircut. Yet the Army, though it can feel at home, is not prestigious, but it does take on an air of reality; the payscale is more closely adjusted to the civilian one and there is a great paucity of employment in the mid-range (between ditch digger and missile engineer), consequently twenty year sergeant is a life that fits into the scheme of things (much as experienced or skilled factory worker fits into the industrialized scheme of the North). Kentucky is commercial and agricultural -- about 60% of the roads are unpaved (on the level) and serve farm areas. The town structure is adjusted to the farm with co-ops, seed emporiums, old west barrooms and hardware stores. A certified first year teacher in the county school system is happy with 5400 dollars a year because in poor counties the figure can be as low as 4800. Ignorance and provinciality are the pride of Kentucky -- They're about all that can be exported and all that can be nurtured at home.

 Michigan with the industrial hub of Detroit chooses to be a northern state. While Ohio and Indiana choose to be mid-western ones. Kentucky is, by its own volition, a southern state, and a fine example it is.

Am I hyper-critical or does it just seem that everyone here would give his back teeth to get to New York City (without having to give up the Chivalay Impala and its stars and bars front license plate). What a sad state to preserve its rights so well and laugh so long at hair half as lengthy.

    Criticism, like the above, veils the true motivation of the individual soldier. He knocks the system, his commanding officer, his assignment just because he realizes that he cannot really talk about the one thing that's keeping him going -- his plans for what he'll do when he gets out, all the things that he can reasonably love. Hate is a much more acceptable emotion for an emotionless army because it is so hollow and so easily overcome. Love is full and a miserable thing to destroy. Better it be secret. Better yet it be forgotten to be pulled on like levis when the end comes. The army can't really mess up your plans unless you allow them to control a variable. Since they have you and, physically, you must enter some of your own planning, there must be a part of your mind that they can't ever get at and it must be a seat of metaphysical plans. Very difficult thing to understand until you've been it and had to crowd the good life away, cram-packed into cubic centimeters of brain unfamiliar...

    I think about a complete life and a crazy house built of brick and steel and stone growing out of the ground and a stable and ring off in well trailed country and time to spend in it all; long, hard beaches to ride with water to do up around your ankles and slide back off; long Saturday mornings with a coffee pot by the bed; some kind of self-employment like weaving baskets or sifting small quantities of sand; time to read all of E. Hemingway over again; and time to sit and look at you sleep; blue sheets to go with your eyes; and royal blue china with a sixteenth inch rim of gold; egg rolls; a huge weird chess set to impose on the living room floor. Who said I couldn't dream?

★ ★ ★

...When I'd just dropped off, an orderly came for me, woke me proper with a flashlight, and sent me off to the orderly room... I was to drive in the morning, that was what they had to tell me... Checked out a 5-ton Chevy stake truck at six and drove it back to company to be loaded with range materials. Picked up my rifle, put it in the cab and drove out to Heins Record Range -- six miles rattling and thumping. It was a 1967 truck with a tight four speed but with the eternal rattle of stake trucks. I was placed in the first order of shooters and fired around 7:30. I qualified so well that I did not really need tomorrows session for final qualification -- if I make eleven hits tomorrow they'll give me a sharpshooter medal. When I was nine years old my father gave me my grandfathers shooting medals -- a set of pistol experts ribbons and bars. Reginald Austin Bailey was in something like the 17th Massachusetts Cavalry, the army in riding breeches and boots or spiral-wrap puttees. I would like to send my father a sharpshooter medal to indicate to him that he can take the pride in me that he takes in his father. Shooting means a lot to Dad and I'm trying mostly for him and doing o.k. because of the familiarity that his rifles gave me with shooting.

It's shooting , Maris, not killing, but I dislike it for it is so much oriented to cutting "Charlie" down before he gets to you. A hit on the silhouette target, which falls when hit, is a "kill"; they write "k" on your scorecard in the proper column. Just as we all scream "KILL" when we jab the air with our bayonets or cut with our hands and feet in hand-to-hand combat. I'll shoot for my fathers pride and do the bayonet and hand-to-hand just to pass the test and avoid a recycle... I can't handle the "kill" stuff and I don't know about it -- can I, when I must, shoot or stab or lash out to save my skin? Say I'm sorry; say pardon me; duck; chicken out -- ? Oh, Maris, I have to assume that Vietnam is waiting for me, especially after jungle training at Polk, and I don't know about the "kill" stuff -- preserve my sensitivity and love or save my poor ass; that's the choice, it seems. I do not want to face that choice, or even be prepared to make that choice. It will come...

★ ★ ★

...The training is, of course, the worst bit, but its all oriented to mounting an effort in "the Nam" and peace is not attainable by this kind of effort. The tanks rumble by for hours and hundreds of us jab the air with bayonets in the heat and humidity -- "it gets up to 110 degrees daily in the Nam" -- "Charlie wouldn't let you say 'please' soldier." No, this I have to work against -- it will have to be from within... it's a subtle, not at all obvious thing about fighting from the inside against the army. It is first a challenge to maintain your attitude and your values -- but that is a fight to keep your head above water only. As a soldier with a moral sense one is still only a potential danger to them. Another gain in potential is to determine where it hurts them, to prod -- for example: it hurts many dedicated officers to be reminded that they are fighting in Nam for containment not victory -- that's not a noble aim in their minds. Or, another example to argue with anyone that the U.S.A. lost the second world war -- it's quite amazing to find out how many of these people think that we won the war -- or can win any war. Soldiers have a conception of victory which is all tied up with the single success of killing an enemy soldier -- if you kill him, he won't fight back, and more broadly, if you kill them all, there won't be any more fight. They can have the fallacy explained to them.

    No. Not Charlies back door, but destruction from inside. I don't want to be abandoned by a group of suddenly moral soldiers on a jungle battlefield. Scary. But I want to make soldiers aware that the concept which rules the men who command the men who command the individual soldiers is dead wrong. And to make them aware that it is their responsibility to modify that conception when and where they can. Oh, it sounds so grand, my war of attitude. But soldiers trained are angry men, not merciful ones; men with a built in over-kill. Cancel that attitude and cancel it in the men on top and you and I and all of them could be on the way to a real victory...

★ ★ ★

...David sent me a very short note to the effect that he would return my uniform, boots, and street map of Warsaw after his mission. He signed the five lines 'Parnell' -- the Irish revolutionary leader, if I recall my Joyce clearly. I sent him back a couple of pages of story like stuff ... A typical David and Ring thing...

★ ★ ★

...Even as I write these words,
 Ruby, I can hear the pulse of heavy armor engines, perhaps less than a mile away, and I fear that you did not get from Warsaw to Krakow within your schedule. My only hope, if that is the case, is that you will be interned rather than the other -- what happens to known agents. I shall have gladly lost a uniform and boots if they serve toward your internment. And, perhaps, they will be a partial payment toward the unpleasant occasion when I destroyed your Lagonda on the Liege road. Those were happier times, before this all broke open, as it had to.
 The ninety millimetre guns crack like huge rifles now -- and last night star shells clustered in the air and big shells beat the ground. Shortly the shells will be on us and the armor clank in the streets -- tonight the parachute flares will light these words. If only you accomplished the vital and impossible in Warsaw -- my map, I fear, was outdated -- can I do more to increase your disadvantage?

★ ★ ★

    This garrison will suffer, for it is a pocket which they cannot afford to leave in any state but powerless, behind them, if I correctly judge their objective. If we keep our heads down and shoot true to the mark,... but that will not stand to their concerted effort. I can call on dated Lancaster cars, mounted with two and a half pounders, but you can imagine them facing off against ten point five centimetre guns as well as I. All is lost, unless hope against hope, you made your mark in Warsaw.
    I saw that the nine millimetre Walther was gone from its holster -- I trust that it found its way to your pocket -- and the pocket flask in the uniform, I do not recall now whether I left it full. Amongst the Poles a man and an officer can do with a bit of drink.
    I have a few Gold labels and fifty woodbines left. They must last the night. In the morning I can hope to be smoking their substitute for tobacco -- without my sabre and my garrison. To defend to the death has a noble ring to it, proud -- but this garrison cannot hold itself until decimation. Fifty troopers with Lee-Metfords and the cars will not deserve their full force. We might prick them well, once they have passed, but to their front we are but another land torpedo to deflect and eradicate. We will meet them and hold while we can; the men will be sorry to have to quit their fight, but I will have to choose the moment well and then give up my side arms.
    There will be another open road and another Hispano-Suiza to match when this is all over. A

gentleman knows that and lives his wars with it. A gentleman knows when his cause is just and also when his war is over. For it is better to live on ones feet than to die on ones knees, behind our parody parapet with a hot Siddley revolver his only companion. You know the loneliness of Poland where each face is untrusting -- so the tanks will look to me advancing and hard shadowed by the flares, when the night comes.

...Alas, there is no quick ending, nor a gentle one -- I must adjust, in part, in part only. I hope... with McNeill and O'Connell beside me at the embattled parapet, can I be but brave?

    A soldiers regards,
    O'Hern

★ ★ ★

... I've been away for four months and it's all mental for me, no physical reminders left. I've nothing to give. I've become simple and unappreciative through lack of practice. I'll come home and accept your gift of a lovely life not knowing why or how or what I am to do to fit in. Your letter explains it all historically, for me that's lost. The country/western song on the radio says "I was born to be with you". and that's true, but it's not working out. Take the lives out of mesh, deny the new insights, brutally snap off the communication and, God, it's hard to retain the sense of beauty and the knowledge of growing love. A short course in the annihilation of the hopes on which a life is foundationed. A demonstration of the power of anti-every-thing-we-know...

★ ★ ★

... is there some dam respite somewhere, longer than maybe two weeks. Bigger than "I have an obligation" and "an inheritance must be re-earned to have value" -- maybe. How many times do I have to run this through to believe that maybe I really have a debt that I owe for the guy with the medical deferment and myself. Or maybe just for him. It's late at night, I'm dazed, but it all seems so unChristly; it has and it will. If I'm really working off my freedom then it's right and good that no one care when I don't make it across that next rice paddy -- but that seems wrong to me, I don't feel it that way. That all does not jibe with the code of conduct -- probably treason or malingering.

...Hold together and remember the boy Loring,
per usual beggarly request...

★ ★ ★

...I've given organized religion another fling of late -- I figured that if military chaplains didn't take a down-to-earth and realistic stance on the God that we all share and share alike, then nobody could. Well, as it turns out, the run of chaplains here is just not to my taste -- they don't seem to comprehend the basics of (a) faith and (b) Love, and they're just like all the rest on applied Christianity. Now for a hell of a long time I was close to organized religion ... The whole thing was too easy -- I did it and never thought of what religion might mean or what those heavy, heavy words might signify to me. Only when I quit did I begin to consider it, and I'm not through my consideration, not ready to go back to the easy fulfilling of Sunday 'obligations', mob confession and mob affirmation of belief, not ready to get off a first name basis with my God and leave him, vested brightly, with a handshake at the gothic doorway at 12:15 every Sunday. What else is there to think about at Fort Polk, except God and freedom and occasionally, the M-16 machine gun?...

★ ★ ★

...Today, when the mail was passed out late in the apres-midi, I remembered that it was my birthday; at 23, an old man, all day going b-b-b-bang with a patented cong bonger. Two years ago, college acquaintance Maxim and I celebrated the birthday with a fish dinner and a series of Michelobs at Hartfords good 'Honiss's' Restaurant -- last year Maris and I bashed a split of champagne, some Harvey's Bristol Cream, and a plate of decent sword -- this year, just b-b-b-b-bang (good six to nine round bursts, men!) and my mind is going the same way: downrange at 2840 fps. October 24 and I've been continually at odds with this organization for four whole months -- and they're beginning to win. Picture me drenched to the skin, pack and canteen, pack suspenders with ammunition pouches, camouflage-covered steel helmet dripping casually around its perimeter, boots squishing, M-60 slung on the hip, a long belt of orange tipped cartridges draped bandolier-fashion over my shoulder, mud to the knees, yelling my bloody off (let's hear some noise, men!), squirting good six to nine round bursts, left and right, running pell mell, raining like hell. The odds are with them, but it's a bold contest Davey -- battles of liberation tend to be that way.

...I broke the communication (via telephone) barrier tonight -- called and talked to Maris for the first time in a good month -- we've been writing for so long (and that's a distinct style of communication) that we couldn't talk to one another. Very unsatisfying ... What's there to say? What are we going to do, leave notes around for one another -- "dust the coffee table", "take out the garbage", "I love you", "turn on the T.V."? What do you expect from four months of inhumanity, instant and total recovery? Pvt. "Bonger" Bailey? His grim, begrimed visage, his light machine gun, and his infinite capability for love. "Bailey, have you ever been in a fist fight?" "No, sir." Hell, I've never even gone bowling. B-b-b-b-bang. Beautiful. And then after the mud wading and carrying on, two hours dripping dry, trousers rubbing like a wet hawser at my crotch and entrenching tool banging my ass, cleaning said 23 pounds of pure hell to a pinnacle of cleanliness normally seen only in concours Bentleys, vintage 1927. Naturally a man enlarges his ability to sincerely dislike. Why, a man could learn to hate, doing push-ups in the mud and cleaning his unified arsenal before he gives a thought to his own slightest comfort. Like hell. Beg pardon, sergeant. B-b-b-bang, sergeant.

...Only an eccentric would run around screaming with a light machine gun under his arm. Granted. Only an eccentric could enjoy it, let it roll off his back, not react to it, relax and carry on a normal conversation with his fiance a matter of hours later. There is no single definition of normalcy -- it's a statistically derived momentary constant which encompasses the infinitude of behavior. To convince myself I fit in, again and again. Let them Bertrand Russell the symbols around and sweetly suck their vitality from corpses -- there's such a thing as "more nearly normal", I suppose. The contest of liberation is avoiding slash scars in getting back there.

★ ★ ★

...You just try to hold your emotions together inside, fight the loneliness, and on the outside retain a military bearing and keep trying, making it look like you care ... Amazing, really, that I can live at all this way with nothing but what's inside my bare and non-olive drab skin to set me apart from my fellows...

★ ★ ★

    I received the big envelope of preparedness reports today -- you know I've been playing chess with individual grenadiers and a-r men so long that I'd sort of forgotten the big picture, forgotten that the US of A was in massive retaliation competition with the USSR. Interesting to read; it is all logical and cold hard facts, unrelenting numbers; you accept it because it's such a firm actuality. But when you back off a bit, Good Lord, that about 4,000 ICBM's that us and them have on call is certainly enough raw weaponry to eliminate civilization as we know it, if not life itself. The concept of massive retaliation sort of was forgotten in the big rush to field brush-fire war forces, COIN (counter-insurgency) troops in 1960, then special forces (green berets) in '63-'64, now regular army troops. RVN -- trained. As H.G. will attest, the bomb shelter, indicator of public awareness of the massive retaliation capability, has all but disappeared. But still more minuteman sites must be constructed to keep pace with the rate of numerical increase of the USSR's launch sites. All of my work on Polaris boats never made me particularly aware of the ever expanding massive retaliation power that we have. Strange.
    Anyway, Boston's museum of fine arts, the MTA, Norm Flayderman's establishment, every Bugatti here and abroad, Disneyland, the local supermarket, the Smithsonian Institution, Kretzer's, every vestige of civilization, can still effectively be swept away by a sequence of phone calls and a bunch of

electrical and mechanical operations simultaneously performed at hidden missile sites. I'm being melodramatic. But I'm poorly prepared to face life without those doubtlessly strategically valuable objects and institutions cited above.

★ ★ ★

It seems that voting the choice between Nixon and Humphrey has little to do with moving toward a reduction in retaliatory power. It's institutionalized -- that's why you so seldom think of it and never question it. We'll begin to consider a means to reduce mutual holdings of massive retaliation weapons when and if the rest of our military concerns are taken care of -- that seems to be the commonly adopted thinking. Doing something personal such as declaring yourself a conscientious objector (refusing to bear arms) doesn't seem to be to effective either. (Quakers and the like who can demonstrate a creed which denies them the privilege of fighting, rifle in hand, are assigned as unarmed medics in the war zone.) There's no means that you can effectively put to use to decelerate, stop, and reverse the rate of massive retaliatory power build up. Ban the Bomb protests, I do not personally feel to be effective; others would disagree with me. But when was the last time you heard about a good knock-down-drag-out ban the bomb protest, 1959, 1962? Bertrand Russell where are you? You seem to accept these preparedness reports with an interest in evaluation and a normal sort of "must this go on" grimace. What can you do? Vote against military appropriations and the expenses for the Vietnam war and the technically-valuable efforts of NASA will be cut back before the m-r budget will be touched, because m-r is a convention and a necessity. Massive retaliatory are only a check and balance system, they'll never be used. Okay. Maybe. But for now it's an awfully expensive white elephant and potentially a good deal more hazardous to life than cigarette-induced respiratory ills.

As of this moment I'm down on atomic bombs. It will pass off, I'm sure. But as long as I'm momentarily enthusiastic -- is there anything one can do? Don't feel guilty about contributing to the welfare of m-r equipped submarines, that's not it. Both for now and on a long term, something must jar the conventionality, the genteel acceptability of ever-increasing hoards of nuclear armed missiles. Wouldn't a bit of advertising, reminding people of the arms race go a long way?

Talking like this is no way to get a promotion...

★ ★ ★

...Much more to the point, I think, is this damned unhappiness. You know me when I'm morose and when I can imitate flies and think of puns. I am now very seldom in that latter state and I'm afraid that I'm losing it completely. People lose their capacity to be independently happy all the time, I know, look at alcoholics and addicts. H.C. says that he is afraid of losing his capability for childishness, not just buying the odd 45 rpm record or matchbox toy, but childishness of outlook, call it innocence, not in behavior, but in attitude. I guess that's what I'm fearfully conscious of now. This army is a dream wrecker -- I can't believe in anything but the animal-like state that I have been reduced to by people who excel me only in swinging a bayoneted rifle around and performing left faces... It seems that no other world can exist... To be owned, not by choice, is pretty terrifying -- it wipes out belief...

★ ★ ★

...Please try to remember me through the periods when it's very hard for me to write. Try to keep an image of me alive; when you go out riding, try to recall me standing by the gate smiling as you pass me, your hair silkily flagging behind you. Try to recall me floor-sitting, back when I used to smoke and keep an iced coke going; try to keep me alive out there. I'm not a great cookie receptacle, way far from Hartford, I'm the same old Ring with complications -- that's what I'd like to remember and that's what I'd like remembered.

My tactical officer has a full grown, still young German Shepherd, black and sleek, a perfect animal in proportion and muscle. The dog follows him and dashes off and comes back in huge bounds and a tremendously fast run. The dog is so beautiful and so free; it kills me. I'm not too beautiful, but once I was free. I want to be remembered bounding and running beyond control. Circling around a nucleus of you with my eyes streaming from free speed and my hair back in flopping bounds. Amazing simile, I guess. I want my freedom back -- I'm sort of earning it -- but so many are free without my expenditure...

★ ★ ★

    The humidity has broken with a downgushing storm which is a contrast of reality to the ethereal quality of my existence. All the leaves in full summer green are dripping and coursing with water. The humidity still lingers with me. My anger and I are the only ones in my little divorced world.

WAR

Seattle,
... You can imagine the prevailing attitude in a compound full of soldiers departing for the wars. Life depends upon three or four hours a day of standing while a loudspeaker blares names, one of which might be yours. The Washington weather just fits the attitude to a tee: large mud puddles connected by fog, a constant, wind driven rain, a very fine rain from skies the color of a bad lead casting. You begin to lose faith in justice early in an army career, here you lose belief in the sun...

If our purpose in this world is to make up a workable life-style, then we'd better begin revisions to compensate for these indecisive bits. Bits where seeking a modicum of joy in all things falls flat on its face. Honest, I look for joy here, within myself and around me. The joy of a cutting rain in the face, of holding a warm cup of coffee, or feeling your muscles when you walk, and putting your head in the crack of your arm to go to sleep. Something so big and so pervasive, the joys pale in its presence, is sitting on my shoulder and its brother is sitting on yours, yelling, "You can't be sure now, can you?"

How strange it all is...

Assemble it for yourself. Take a developed case of fear of death by warfare, add a protracted homesick-child-at-camp syndrome, shake in a wobbly set of social morals and a sketchy set of personal ones, season with a big brother paranoia and a liberal dash of shakily reconstructed self-confidence. It's a bloody wonder that

I'm still ambulatory. Nobody headed off to the wars, is exactly in the psychological pink of condition -- but this is a bit much.

Remember the 22nd of June 1970 only and go on, looking for a modicum of joy.

★ ★ ★

    Today… I know that my parents are going to be with you and that you'll have a fairly pleasant time, a good meal of stuffed grape leaves. Sympathetic and unsympathetic people are both problems. But I think that my parents most nearly share your own attitude; things should be easy. Today, as I go away once again, I'd like to think that things are not too hard for you. I love you, and as a son, I love my parents too, and that's about the entire spread of my emotion. Today I can project it all toward one geographical spot.
    I have been called to go.

★ ★ ★

...Cam Ranh is very sandy, being a port facility, it is literally the beach hemmed-in by brown mountains on two and one half sides. Cam Ranh is relatively safe for Vietnam, employs a lot of local labor (but not Vietnamese KP.s) and offers all the convenience of a stateside military reservation; PX full of Prell and Krackle bars and Kodak film, snack bar and all -- that and more serving only the 54th Replacement Battalion and, of course, there are many more Army outfits headquartered here, as well as Army aviation (rotary-wing) and Air Force and Navy units and a commercially served airfield. Still, there's no plumbing and every building has a reinforced bunker beside it and the aircraft swoop low and loud all day.

God knows how long I've gone sleepless and there'll be no sleep tonight, I can tell. I've lost track of Pacific Daylight and Vietnam time and flight time and the dateline. I left on Tuesday and arrived on Thursday with no intervening Wednesday. And God I need some sleep. Constant sweat and a boggled mind make Ring not to sure that he won't get morose at the drop of a lousy assignment. For now, though, I'm still in sturdy, good spirits. I haven't heard a shot fired in anger yet. I haven't heard a shot.

It's raining hard on the tin roof of this barracks of two by fours, lattice, screen, and corrugated steel. Tropical construction: open ventilation with long roof overhangs to stop the rain from coming in. The roof leaks...

★ ★ ★

Today I finally got my assignment to a unit...
The Americal Division...

★ ★ ★

All is well in Chu Lai; three more days of training and I'll be going out to my unit. The Headquarters of the 11th is down at Duc Pho and the location of my little part of the 11th is at L.Z.Liz (L.Z. for landing zone), just above Duc Pho.

My personal attitude has taken a little bit of a blow, though. One incident has utterly soured me temporarily... I was paid for August, September and in advance for October... I received a wad of four hundred dollars in MPC's --- several spites worth, (Austin Healey). My wallet was already filled to cramming with nickel and dime notes. So I placed the wad of twenties in my stationary case and locked it in my duffel bag. Safe... In the frenzy of an alert (a siren wails and we all head for a sandbagged bunker in expectation of a 122 mm rocket attack) my bag remained unlocked and some very skillful soul opened the bag, the stationary case, by-passed a Parker jotter, an Exacto knife and steno pad to grab a bit more than four hundred dollars (I'd added a few stray ones from my wallet to unpad my wallet).

So I have about sixteen dollars to last me til November's payday. But it's not so much the being relatively broke, it's the stupid feeling of being a careless sucker... it's hard to prevent the stupid stupid malaise from extending into my thoughts -- sucker for being in Vietnam at all, sucker for being in the Army having life and death decisions made over you by illiterate sergeants, sucker for letting them spoil your life with your wife. It's ruined my otherwise good general attitude.

...I'll get over my loss of attitude sometime. I've done equally stupid things before and lived, I suppose. There must be a moral there somewhere. And all that.

Vietnam is getting me down a bit. Helicopter rides are no fun, I assure you, when you're waiting for holes to happen in the floor. And some washing water would be awfully nice.

Before I upset you (oh, well, I suppose I have) I must assure you that Chu Lai, and even little L.Z.Liz, is pretty safe and I'm in no danger other than psychological -- and that I'm working out telling you about it. Nothing can be done except let the sucker feeling slowly dull down. Talk about it; re-draw the simple moral.

...Simple men, exposed to danger and a touch of privation, express their desperation in spectacular actions. That's the moral. I'm strangely calm, so is English major T.; we continue to joke, to trade books, to even talk authors. Others gamble spectacular money on cards, repeatedly get drunk on beer, sleep with their empty rifles. Not kids, people who've had their beer drunks in basic training (instead of sophomore year at college), people who've been here before. It's a nervous place. Just essentially a nervous place. A touch of self confidence goes a good way.

And I've still got most of mine...

★ ★ ★

...It's the beginning of Monsoon Season (capitals for emphasis) here -- October to March -- there are daily torrential rains which apparently build to intensities unimaginable, then taper off corresponding pretty much to the bad weather in Connecticut. Daily temperatures still reach in the nineties and humidity is quite incredible. Most of the time you're so waterlogged in your own perspiration that a good solid shower isn't noticeable. Average intake is eight to ten salt tablets a day -- a very impressive figure. And malaria pills of course; a little white one every day and a monstrously big pink one every Monday.

Local vegetation, along the sandy coastal plain mind you, runs from evergreen to coconut palm and from cactus in shrub size to elephant grass. The heavy vegetation begins much farther inland than I've been, but I hear double and triple canopy jungle spoken of as a matter of course. It's still a bit hard for me to imagine that I'm in Vietnam -- all I've seen is tropical military installations without hot water, plumbing for the most part, and reliable electricity. I'd hate to judge the U S of A on the basis of Fort Knox and Polk in the wrong season. We'll see much of Vietnam before this thing is out, I'm sure -- that jungle, and civilization too. Never much on tourism with an M-16, I'll have to try to retain my impartiality in the face of prejudice in order to get any kind of valid impression of the place...

★ ★ ★

    I'm at the headquarters area of the good ol' First of the twentieth, "Sykes' Regulars", the battalion hand picked by General Sykes in 1866 from his Civil War-hardened battalions, formed in 1861, to go west with Custer and get their asses kicked at the Little Big Horn. In defiance of hallowed tradition, Company D, 1/20 hasn't had a casualty since mid-August ... Delta Company is the apple of the battalion (Sykes Regulars) Commanders eye, so we're selected to defend his personal fire base (Liz) when the action peaks...
    It's a strange feeling; the war isn't here, it's at home. And if the troops were all withdrawn the issues would still exist, but nobody would bother. My function is to make it imperative that the issues be met. One does hesitate to risk life or limb in pursuit of such menial duties. Carry the seventy pounds of rucksack and the silly fragmentation grenades and the PRC-25 radio, okay, but getting shot at... I'm an issue emphasizer, not a soldier, one each.
    Sitting in the Club Tiki; pieces of about six camouflage tents, some four by eights of plywood, Sampsonite folding furniture, quantities of warm Schlitz... About a hundred and nine degrees in here. Sunburned and dribbling sweat. A quadruple mount of fifty caliber machine guns and a couple of 81 mm mortars, up the face of the hill, going to town all of a sudden...
    The time passes rapidly in retrospect. Time spent on insect repellant and loading magazines, filling sandbags, remembering malaria pills, radio batteries,

C-rations, filling collapsible canteens, reading a bit of Faulkner and some Tom Wolfe. The war is a lot worse in Connecticut. So I hear. I've only got eight months of it to do over here; in upstate New York they've got to live in it forever.

★ ★ ★

...When I write during the day, I soak the lower half of the paper with my arm -- the ink spreads out venously when it hits the damp paper. In the cool of the evening, the light is gone, as is the perspiration and the inclination to do anything but go to sleep. The heat, I think, is quite tiring just by itself, seventy pound rucksack aside. So you'll have to get used to pre-soaked letters and I'm afraid, letters few and far between.

Reading the introduction to Faulkner's The Wild Palms, I thought that it would have a far from liberalizing influence upon my feelings about abortion... It seemed to me to be awfully reminiscent of the Hemingway Paris/Pamplona novel The Sun Also Rises, the central figure of which is incompetent in another way, simple, a little sadder, but just as completely involved with loving. If anything, The Wild Palms showed intolerance for abortion as a result of unwillingness to discard anything of love by people who suspect the love between two people to be much larger than the people -- Something larger than the arithmetic sum of the constituents. Does that make sense? And, conversely, that love which is merely equal to or smaller than the involved people produces results (i.e. babies) which can be little more than hindrances to the people, restrictions upon or reductions of, the finite sum of available love. Then, romantic or realist, choose your weapon, a filthy catheter will do, and take it from there, because Faulkner hasn't really solved anything, just insulted you if you happen to suppose that personal philosophy and the matter of abortion can be at odds. Processes of intellection aside, I remain firm: you may not pierce your ears.

★ ★ ★

    I expected L.Z. Liz to be sort of a clearing hacked out of the jungle -- something not too civilized. In fact, it is the tops of about three low hills rising out of a very pleasant-looking cultivated plain. Heavily fortified, of course; the guns occupy the crest of one hill, a heli-pad the next, and an observation point tops the third hill. Bunkers ring the three hills, about half way down the slopes, making, in-all, something of a figure eight with an extra loop for a perimeter. No, civilization in the form of cold Coca Cola is not available -- and these are hot naked hilltops. Duc Pho is visible from here and on the other side of this hill you can see the South China Sea, a thin dark blue band beyond the coastal piedmont. The 155 guns fire with a hell of a big bang over our heads and you can hear the big shells spinning away, like tiny planes, for miles. A 155 gun can shoot about 22 or 23 miles; I can't see the shells strike from here.

    We expect to be on Liz about two weeks; I'm not sure why so long; we'll be sending out some patrols and ambushes into the immediate vicinity, not more than one or two kilometers. But for the most part we'll just be pulling security around Liz and be on-call as assistance in the field. During the day there are sand-bag filling details, road security, garbage pick-up details. Just like the army...

    I miss you. I'm awfully busy and the time is passing, but the nights without company, out in a cot, watching the tracer bullets from our quadruple fifty calibre machine gun mount whip over are a lonely time.

Sweat and insect repellent. Tapes of radio programs made in Detroit on July twenty-third. I miss you; you understand that...

★ ★ ★

    I ran out of light last night; the twilight became dark and the smoking guns became smoldering dark red patches in the blue black. I stood a couple of hours guard -- and I can say that all was quiet between midnight and one and between four and five. I sometimes want to write an account of every minute and every sight of interest. Then in moments of firmer mind, I decide that you are probably sufficiently upset by my mere presence here and that it doesn't help much to send a neat description of seeing a tank of napalm flip off a jet plane or even of my units patch. We never said much about the hard possibility of my not returning from Vietnam, mostly because I plan to return, and under my own power; but each of the military vignettes in a letter makes that possibility more real for you. On many an occasion I have and I will want to write, but I won't have good material to do it with, sometimes, with pen in hand, even, I just won't be able to write -- but you know how that is. Other times I won't have the opportunity to find out that I can't write. Those will be the hardest times...

★ ★ ★

...I'd like my olive green bush hat, which is hanging on the closet door in the back room, even though it makes me look like a drunken forest ranger or scout leader. I think that the bush hat will be lighter and cooler than my non-bullet stopping steel helmet, and it is authorized wear, in the field only.

...I got an excellently heartening letter from Rik on the moratorium and what is planned to go after it. Would that everyone wore black crepe, not for the thirty-seven thousand lives that are gone, but for the death of this war. I'm pleased that you took part and that the whole thing was so successful. An avid observer on L.Z.Liz is watching the progress of the war on your front.

Last night I went on a 'snake' (that's the word for a small ambush patrol). We set up on a little scrubby tree line about a kilometer-and-a-half from the perimeter at Liz and watched and waited for activity on the access road to Liz which is a favorite target for guerilla mining operations. All night, five of us, three rifles, a machine gun and a grenade launcher between us, waited, watched the shadows move, and smoked furtive cigarettes in cupped palms. Nothing, no action. Just a dreary way to spend a hot, moist night, sitting, listening to your rifle rust, imagining that every bush just slithered toward you a yard or so. I should sleep and sleep some more...

I wish we were together now, somewhere, talking softly about most anything, knowing that we had an envelope of love around us, visibly radiant and awfully mysterious, as warm as you and as hopeful as me. Eight months, now. Two to our anniversary, two more to our R & R, then a downhill slide to June thirteenth, and another summer right on top of the one we just had; getting used to one another and snapping out of it, movies again and sleeping in front of Merv Griffin. Coming home from work and calling up at noontime, sticks of candy and art projects. Building a house and eating casseroles; you're nice to love and hard to be away from, better than Dinky Toys and bigger. You must be real.

★ ★ ★

...Vietnam is a very pretty place... (it) could very well be the Riviera of the Orient given air-conditioning and two or three hundred years of modernization. Wooden treadmills, human or oxen powered to pump irrigation water; hand threshing of rice, in sheaves, into baskets; sophistication is, mechanically, the sight of a so-called 'Lambretta' trike (mostly Japanese made) like the mail used to travel in at Electric Boat, with about six people hanging on the cab and conestoga covered bed, and the whole thing streaming with more flags and bunting than a Sicilian donkey cart. I haven't seen a real car yet, a few Kawasaki busses and an occasional Toyota truck bodied like a Volkswagen bus, sometimes an old old International Harvester truck chassis with a combination gypsy-conestoga box body on it, painted sky blue (more often than not) and chugging painfully. Universal is the fifty cc motorbike, moped and clumsily motorized bicycle; four people, three of them side saddle, on a Honda, the whole thing making maybe three knots under full throttle in second.

Wherever the army is located in this bright green country, there are scabs of raw, bulldozed and eroded earth, smoldering junk heaps, and rusty culvert sections, rust and dust and corruption, broken lumber, and sagging concertina of brown barbed wire. I'm really ashamed of it. Never in the past could I ever have thought that the parallel between this conflict and the American Revolutionary War was so close, and I suspected it before. We're really having asses made of ourselves and paying well to have it done.

    The bombing halt? Perhaps those multiple contrails that I see every night, headed out to sea above the broken clouds (I thought, for a while there, I'd adjusted to the gunfire) aren't what I know they are. Perhaps it takes a flight of three B-52 bombers to reconnoiter Cambodia and the DMZ? Ever wonder why the Paris talks are so stagnant when we seem to be conceding so much?

    Enough of this, I'm not that bitter, just sorry we have to spend a shooting war on this. Yes, I'm being careful, very careful. Yes I'm working on getting a job, clerical type, in the rear. Right now I'm awfully bushed from the 'snake' last night and I'm sort of disorganized, but I've wanted to write to say thanks for the letters, a big morale boost...

★ ★ ★

...The rain is now perpetual, but things are manageable once you adjust to being soaked. Liz is like a newsreel of the battle of the Marne; everyone slogging around in knee deep mud and trucks bogged down to their differential cases. I just don't want another snake tonight...

There are a couple of things I could use here;... Food stuffs, although they'd be nice, aren't reasonable because of spoilage and the fact that I have to carry my material possessions wholly on my back -- even a couple of pounds of good cookies would be a bit more thump in the small of my back when I exit the hovering helicopter. I would like a small steel mirror, for shaving when there's water to spare, and I'd like a decent knife, my much abused, Swiss Army jackknife isn't up to the rugged tasks... M-16 bayonets are not an item of issue (well, can you imagine doing a vertical butt stroke and slash with a pistol-gripped rifle?) and decent rope cutting, wood gnawing, bamboo splitting knives are scarce in the platoon...

I can't offer too much in the way of local products in return. Motley crews of Vietnamese waifs purvey marijuana rolled into cigarettes, sandals cut from pieces of tire tread and slices of innertube, peace medallions and switchblades made in Japan, strings of glass beads, and bracelets made of aluminum tubing. Thus far I've resisted buying. If you'd like some Ho Chi Minh racing sandals or a packet of reefers, let me know...

★ ★ ★

...The local marijuana is, for the most part, doctored with opiates to make it quite a different thing from Mexican marijuana. It is distinctly stronger and more sleep-inducing, or more passive, although the characteristics of the marijuana high are still there. Marijuana is available both bulk and rolled into joints, at not much greater cost; very popular are about four inch long joints called 101's which go for about a dollar and a half (American equivalent) for ten or a dozen. Which is, of course, ridiculously cheap by what I guess current stateside costs to be...

★ ★ ★

...The policy in the battalion is that the four line companies each spend three weeks in the field, then one week at Liz, then after three cycles, each company goes to a stand down, three days of beer and real bunks in the relative security of divisional headquarters up at Chu Lai. Now the exception which proves the rule. Alpha and Bravo Companies are on semi-permanent assignments away from Liz, so we are to go into a slow cycle rotation with Charlie Company; they'll be up here for, perhaps, six weeks after which, I expect, they'll go back to the field in our place while we go on our stand down. How much of the foregoing is rumor, I don't know; where we're going is still more suspect as rumor. Anyway, I'm going out into the monsoon where stationary gets wet and experience is monotonous and I'll get to feeling rotten about not replying to good letters and sleeping in wet clothes will get passe'...

Today word of Nixon's 'hold the line' policy speech on Vietnam became knowledge here. Apparently Moratorium Day was not too effectual on policy. The continuing program of demonstrations will have to either be firmer or actually disrupt in order to make disfavor more immediate to Nixon. He wants to save troop withdrawals, a plum to be dropped immediately before he attempts his re-election. Too bloody much politics have been made of this war, at the cost of 37,000 lives and untold monies usable on internal problems. I'm afraid that firmer demonstration will lead to schism over 'violence' and ferment... Taking it to Washington, literally has been done, Armies of the Night and the Pentagon thing have been done. Magnitude rather than individual spirit is necessary -- ward healers rather than masterminds.

★ ★ ★

At times I wish that I carried an instamatic-type camera in an ammunition pouch as a lot of people here do. I'd like to take a picture of Duckly, the duckling, whom I've been carrying in my helmet for a couple of days now. I saved Duckly from the swinging Collins machete of our point man who considers all winged, but non-flying things to be fit subjects for his trail clearing operations. We walked and walked and then Duckly and I got CA'ed on a helicopter into the mountains. Duckly is sort of a filthy yellow color and he grooves on cubes of freeze-dried potatoes softened in reconstituted milk. That's a taste I can't develop. Duckly and I then stumbled and fell down the mountain all rugged and rolling rock heavily vegetated; machete work, took all day. Forehead and nose slippery sweaty, the vines catching at that three pound steel helmet and slamming it back down on top of glasses, bridge of the nose sore and glasses on the tip of the nose, Duckly in hand. Then Duckly and I rode another helicopter, a big Chinook, and another doorgunner looked askance at the infantryman holding his rifle and the tip of his radio aerial and a yellow duck. Bought a Pepsi-Cola from a Vietnamese waif, fifty cents, going rate, and Duckly and I walked and walked again. I had fifty cent piece sized blisters on my heels from going down the mountain in wet boots and every time Duckly made his pitiful little squeak I agreed with him; the vines and brambles would catch on my radio and rucksack and inertia would take me for a giant bathic somersault into the elephant grass. I'd flip off my pack like a flying disjoint on the rings as I somersaulted and I was afraid that Duckly, in my hand or my helmet, would lose an eye or get cut in the grass that slashed up my forearms and made my rubberized-canvas jacket into a fantastic collection of three-cornered tears. Duckly made it intact. Today I fed him up well on my lerp ration and, since I was on the end of the line of march, I let him go on a rice paddy dike. I figured that he was safe from machetes and he's still too small to be a Vietnamese dinner. He was a proud duck... he only submitted to handling. He didn't look back. Maybe his ears still hurt from the helicopter rides.

★ ★ ★

Just inland of the 'Gaza' strip there's a river. In the paddyland adjacent to the east bank of the river there are several small villes, each built on man-made elevations of the land to prevent inundation when the monsoon flooding comes. Like now, for example. We waded out to this abandoned ville in waist deep, whoops, chest deep water. The land between the river and the limits of the bulldozing on the strip are a free-fire zone --we can shoot at anything that moves because all the civilian are supposed to have moved out to refugee concentration camps... The ville is practically all rice cache; under the floors of all the hooches are buried 55-gallon drums capped with pliofilm, of rice; maybe ten or twelve ton in all. And there are about four families of pigs on the island, from eight inch piglets to hardly mobile sows, and chickens from fuzz ball chicks, through gawky chickenlets, to full plumed roosters. And ducks and rats and lizards. The people who have lived in this region try to sneak back in to recover their pigs and their rice and their former selves. We get some sniping -- some AK -47 automatic fire from the woodlines -- so everyone's scared and return fire is overgenerous. The lightless, furtive life in the burned villes is doubly dangerous because of our machine guns and the artillery and armed helicopters we call in. Six civilian have been KIA and we've evacuated ten or a dozen with burns and shrapnel wounds. We've captured and sent on about seventy five of them, women and babes in arms, elderly and limping. Still we get fired on; I had the unique experience of being fired on by an odd American BAR. When we'd run off the sniper I found his expended cartridges; they were made at Lake City in 1966. Yesterday mortar rounds came in. Some of this rice is going to support a fairly well organized VC operation, but only civilians die. From the rear S-2 sends out word that some of the civilian types we sent in are VC-sympathizer-types and auxiliary organization leaders. So we're safe, the VC stays in the meager scrub on the strip and we sit in fat city on our island, bagging and shuttling out rice on helicopters. And the civilians get in the middle and get shot up. When we sweep out of the ville, across and down the strip the VC and the weapons are gone and only the infirm clinging to baskets of rice and dried rounds of potato remain.

    Who really cares who wins as long as the civilians stop getting messed with and having their rice taken away. Perhaps, as Radio Peking says, "Raising high the great red banner of Mao Tse Tung thought..." The Chinese see Vietnam as an agricultural resource to add to their bloc. Perhaps, even if we left and made the VC unnecessary, someone would continue to take the rice away. I don't think that we can economically elevate these people to a point where socialism is not inevitable. We're not really trying. We give away some new Dodge trucks (for export vehicles, the old name, Desoto, has been reincarnated) and buy a lot of Pepsi Cola at outrageous prices. To move the whole economy up a notch or two some regulation is necessary, maybe even the "great red banner of Mao Tse Tung thought" is the, or one of the, alternatives. I think there'd be less shrapnel wounds if not less starvation. To a staunch capitalist, I'm the worst sort of Commy-symp, but Vietnam is not mud-fifties America, nor is it even

England or Sweden where measured socialism seems to work. Picture famine as a way of life.

Maris, please excuse these pages which result from an attempt to rectify the visions of bloodied people and incoming fire. We've had two people killed and a half dozen wounded with fragments since we've been in the field. People I know, who came in with me; I talk in the med-evac helicopters and send up the medical details. For aggressors, we're certainly taking it on the chin. We have a small ARVN element with us, seven national Police types, your smiling Gestapo, with revolvers and carbines and a whole bag of torture techniques. When they don't have any business, they shoot pigs and stomp chickens. What the hell, so do our people. In a war nobodies right, or half right, better or worse, I guess; the National Policeman shooting a civilian in the hand or foot, me sticking to my radio and not shooting, a sniper believing in Nationalism, or a Washington peace-marcher overturning a car, even. Some people believe they're right and act on it and others don't know or care and act like those who know or think they know, because they have to or want to or are too scared not to and everybody's wrong. Because they're making somebody do something. More guilt complexes I don't need.

After a week in the field that's what you find filling your head. Reading back, it's all cliche. I've heard it before and to have it written softens it; takes the sound and smell away. It's a personal sort of experience, you have to either write all of it or none of it; suffering as it does from rumination and condensation.

★ ★ ★

next day

I have a rope hammock slung in a hooch, a perfect woven net, hemp hammock. It gives you the lower back pains that a Sealy Posturpedic is supposed to prevent, but it's neat. Last night the platoon sergeant was crawling, stooping under the hammock when I woke from a nightmare. I kicked him twice in the kidneys, swung and knocked him square on his butt, then woke up and asked him what was going on. Waking, I'm passive enough to warrant violent dreams or upset enough to waken pugnacious. I kicked three strings out of my hammock and worried the rest of the night that it would rupture -- about forty strings make the net. Today I sewed and bound and whipped off on the sewing and I've pretty much regained structural integrity. That's a more reasonable explanation of war.

★ ★ ★

    While listening to the radio reports on ARVN of the more "apprehensive encounters" on Connecticut Avenue where a police motorcycle is reported to be in flames and federal troops are billeted in the basements of the grey stone buildings...
    We've set up here in a little ville in the flooded paddies adjacent to a river. It's an island surrounded by waist deep paddies and knee deep paddy dikes. We're turning up rice caches in the floors of all the hutches (shack houses) and flying it into our forward fire base on a helicopter shuttle. A couple of times a day we get some sniper fire and we shoot back at them and call in artillery and, last night, Cobra gunships with 2.75 rockets and mini guns. This is a free fire zone, about four or five miles from the beach, all the civilians (neutral Vietnamese -- technical term) are suppose to be out and anyone remaining is either VC or VC sympathizer. Of course, there are people who live here, have lived here all their lives and won't go to a refugee concentration camp (New Start Camp -- isn't that a Big Brotherism?) just because we declare 'free fire'. This is an abandoned ville, but in some of the other villes (one of them in visual range), furtive life goes on; the people sneak in at night trying to recover their rice, their pigs (this island is a barnyard of pigs and chicks and ducks and rats and lizards -- getting along with amazing amity) and their sense of self. The machine gun team shoots a lot of them in sniper scares and artillery air bursts (which are horrifyingly effective) gets a lot of them. We've dusted off (helicopter medical evacuation

or 'med-evac') about ten of them in two days, killed three or four. No sweat, free fire zone, everybody's a VCS. I still haven't fired this rifle --

The last time I saw Dupont Circle it reeked of pot and propositions from the wrong gender and general co-existence. That was August, headed up out of downtown's heat to Georgetown's wealthy cool. Right then there weren't any confrontations at the Pentagon, no fire, or, the radio says, troops and overturned cars. Right then I wasn't too worried. Damn fool. How you gonna act, Dicky-Babe? Said through angelically gritted teeth. Ginga Nai means plastic bag in Vietnamese. How does one act? Along Connecticut Avenue? Only two kilometers from An Tho? Down on South Main Street? When we switch the radio over to short wave we'll get what the BBC Far East Net on the eleven meter band has to say and then some hard core, 1955 style commie propaganda from Malaysia, "The Nixon Administration in power less than a year, is already at the end of its tether". the economy is always tottering and all figures released are always watered down. The Malaysians may very well have it straight. Then Hanoi Hanna will give us some super subtle-psychological rapping, but we have to wait for Hanoi to back into the BBC's frequency. Everybody will talk about Washington. "Imperialist Chattels." I have a rope hammock slung in an empty hutch, my radio's hand-set and an L-head flashlight and a couple of smoke grenades (a violet and a green) hung in the hammock. One "pops smoke" to bring in helicopters: "Dust-off-eight-one, this is six-two-tango; I have smoke out, over." "Roger that... I see your electric chicken (or goofy grape)." "This is six -two-tango, roger; come on in." "This is dust-off-eight-one-medic. How many litter cases do you have?" "This is six-two-tango. Wait one... We have two Mama-sans and one baby-san who aren't ambulatory cases, over." "Dust-off-eight-one-medic. Roger that. No more than two to a litter please." "Dust-off-eight-one, this is six-two-tango. Are you an Imperialist Chattel?" My hammock is rocking. BBC will come on for evening programming in fifteen minutes. The light is failing. Three big chicks, their heads bobbing like Don Martin cartoon characters scoot by the door. And Arthur, the little pig with the pink tummy. The pigs have longish, non-curly tails and they wag them all the time; they seem so friendly...

★ ★ ★

...I wish I could pinch a couple of five ton (off the road capacity) dump trucks and a front end loader for you with the ease that I've come by a rubber poncho for Maris and rifle rods for my father. If, however, you receive a rather large package without a Customs Declaration sticker... do paint them yellow as soon as you can, the big star on the door is a bit of a give-away... Perhaps you'd like a smallish parcel without a Customs Declaration -- fifty pounds of C-4 or Composition B? It's never struck you that Santa Claus looks a lot like Ho Chi Minh?...

The primary mode of mechanized transport in Vietnam, with the exception of military deuce-and-a-half trucks, is the Lambro. A Lambro is a trike based on scooter running gear and a 150 or 200 cc engine; They're made by Innocenti-Lambretta in romantic Italy. There's a semi-aerodynamic cab with an inverted-half-cone fender stuck on the front of it; a scooter front fork complete with trailing-arm suspension and steering head, with rectangular speedometer and twist grip throttle, pokes through the fender to serve the driver who has a small bench seat (room for three Vietnamese). The bed is conestoga covered like a miniature VW pick-up and the rear axle is centered under the bed to take most of the load off the front wheel. The drive is by 3/4 inch shaft to an asymmetrically-located gear carrier, which I suspect, doesn't have a speed differential capability. There's a small two-piece, pressed steel muffler with a shape reminiscent of old Renault Dauphine mufflers and the sides of the bed are stamped (like pick-up tailgates) with "Lambro 550", a bit of an exaggeration, although one wonders how 200 cc's manages to haul twelve to fifteen Vietnamese around. They put four (driver straddled, three side saddle) on a Honda 50, though, so maybe...

Traffic goes at a weaving three miles an hour and, for the first time, quarter ton Jeeps look squat and powerful. Now there are a few biggish Dodge trucks (bearing Desoto chrome -- new trucks, so the good old name has been re-vivified for export purposes -- Desoto lives!) of the commercial type running around. I welcomed the sight of commercial trucks as signs of emergent Capitalism the absence of which is this war's guarantee of prolongation and eventual "loss". But I was unhappy to discover the trucks to be gifts rather than outright purchased items. Commercial charity is, I suppose, an improvement on military aid, and mobility of agricultural products will produce some healthy land barons (paddy baron's?). Typical of U.S. foreign aid, though, the trucks all seem to have starred windshields and are rapidly running down. Rather than more trucks, send follow up maintenance, parts, and good lubricants. Not bloody likely. I'm enough of a youthful socialist to admit that not everybody can pull themselves up two hundred years by the bootstraps, but there's such a thing as burdensome assistance.

Enough. As the oriental sun sets gently over the snipers and touches its last golden rays to the olive claymore mines, we bid adieu to Vietnam, land of mystery and mangled civilians.

★ ★ ★

...We were in the field for two weeks, practically to the hour and now the company is back up on Liz for about a week. Immediately before we moved out I was made, without benefit of ARRL coaching, a radio-telephone operator (RTO) and after a couple of days I was the Lieutenants very own RTO, senior of the platoon's RTO's and responsible for daily re-supply and all communication going higher up, custodian of the ciphers, the rosters, and frequency shift data, and above all, proud possessor of a twenty-three pound radio, spare battery, antennas, spare hand set, and a winning way with a push-to-talk switch. Extra weight to carry, but you get to keep up with what's going on in the broader sense. I've tried to record or recall or explain the infantry-combat-face of war thing in some letters but I've torn all the letters up. I'm neither Ernie Pyle, nor Bill Mauldin, nor Hemingway and I'll let their things stand unmodified, -glorified, -limited, -embellished or re-asserted. Nor have I a good, gory, or funny story...

I'd appreciate any word you have or clippings on the Washington activities of November 15-16-17. The encounters seem to have become more apprehensive and our President more adamant and secretive. Perhaps Spiro has unleashed a revealing blooper? The Vietnam Day Committee in California has been active for years, but they've been ineffective -- the west coast is progressive but it doesn't seem to carry a lot of political weight. You can't get a Goldwater in the White House, but you can get a fairly half-assed movie star in the Governor's mansion in Sacramento. Nationally, it's only when Boston and Bangor get on the bandwagon that any response comes from Washington. And Vietnamization goes on rather casually. ARVN units still don't have artillery support, med-evac dust off, gunships, air strikes; although ARVN's for the most part have M-16's and M-60's, and National Police... and Propaganda Teams have M-2 (automatic capability carbines) and their ammo supply is rotten and their fire discipline is non-existent (even worse than ours -- and a basic load is about three hundred rounds -- fourteen twenty-round magazines with eighteen rounds in each of them -- and I make a daily re-supply order, contact or not) -- it will be a lengthy operation to phase us out of this war -- and if we snap it off, the ARVN's will be out of ammunition and jet fuel in three days, with pieces of helicopter all over the countryside. In six years of involvement we might have done a bit better preparation for our withdrawal; that may be more the fault of the Army itself than Washington policy; the Army always aims for establishment and autonomy in the military situation. Blaming the Army is not just a levying of responsibility, for it is, in truth, merely a job contractor upon which control and inspection should be imposed -- but some of the government is intimidated and other parts are under obligation, and that's politics, double-dealing and snow jobs right down the line. But where's my youthful idealism?

    The company will go back to the field on the fourth of December -- after two whole weeks on Liz -- and stay out again until, I expect, Christmas or thereabouts. The time here, passes rapidly, thank goodness, I hope it won't have to pass rapidly for ten more years of G.I.'s...

★ ★ ★

...You know O.Henry's Gift of the Magi, where the wife cuts off her hair and sells it to the wigmaker to buy her husband a watchchain, while he pawns his watch to buy her a set of tortoise-shell combs? No? How about the story of the guy who quits OCS because he can't face his friend but has to go to Vietnam and his friend goes to jail because his friend goes to Vietnam? If O.Henry had written it, it would have been a better balanced story. O. Henry didn't write it, friend, neither did Ernie Pyle, who preferred dead soldiers with Salvation Army testaments in their hands; his version was bloodier and my my, better balanced. I trust they do something more civilized than stamp letters KIA nowadays. Never trust the Army for civilization though. That phrase about there being something better than dying on ones knees. Perhaps it's doing it lying down. An infantryman is ever in search of comfort...

    I had your letters, until this morning, in the bottom half of a Hersheys Tropical Chocolate Bar box and the weight of them made me unable to answer -- so much to reply to after so long a silence. Today I boxed up and sent to the rear all my letters and six of my paperbacks, to lighten my rucksack (we're going to the mountains). Some people can burn their letters and throw away their books. I have a problem there. With my bulk of reference materials gone I thought I'd be able to scream off a real au pont letter and go to the field having given you a decent and understanding letter. I began a decent and understanding one but tore it up and began again at this older and for some reason undiscarded letter.

Perhaps after this tentative return to communication I could write something more of my own. But I'm going to the field again. And I have to say Merry Christmas now because my next re-establishment will honor the new year. Honor, I trust.

In the calmest of frenzy.

★ ★ ★

Today received your latest parcel of goodies -- excellent mail time and package quite intact. I guessed from the size and shape and heft that the contents were in part Corgi -- but, of course, no idea which one. Vast respect for your good taste and for the quality of Corgi's reproduction. The injuries that the Mangusta suffers in my pocket will, I hope, add character, rather than promote destruction...

Daily, literally daily, my mother clips the motor sport column from the New York Times (and, while John S. Radosta is abroad, people like Roger Penske are doing the column) and sends it to me. So the two best sources of well written information in New York are at work now maintaining my modicum of civilization.

No one else in Vietnam, I'm quite confident, can withdraw from his pockets a book of Updike, a soggy wad of newspaper pictures of the Grand Prix of Macao, and a current Corgi catalog. Since I'm the Lieutenant's RTO (radio telephone operator) I can also produce today's cipher, a company roster, a sheet of frequency shift data, a big, folded up air panel (a sheet of rubberized cotton, day-glo orange on one side, cherry red on the other, used to bring in helicopters when popping colored smoke is impractical), a ball point, sunglasses, wallet (casually mildewing in a plastic bag), a key ring sporting an Ecko beer can opener and keys to an XK-140MC Jaguar, a machine loader for an M-16, a little paper wrapped bottle of Benzalkonium Chloride Tincture (Army Iodine) made by B.F.Massengill Company (and, if you recognize that name you've been reading the ads in Mademoiselle too carefully), cigarettes in an aluminum Ronson Typhoon lighter, British Empire Made (which is the nice way to say Hong Kong). I'm sure you wanted to know what's in all those pockets on the 'complete soldier'. You've contributed not only culture and sophistication but interest and vast distinction to my pockets...

...Perhaps you've seen something in the papers about a 'Village Massacre' committed about a year and a half ago by our sister company... It's a hot story, being the most significant massacre since Sand Creek (Sand Creek: 128 squaws and children, My Lai Village: 109 men, women, and children on March 16, 1968), and the place is crawling with Time magazine reporters. Massacre of civilians is no novelty, really, what with an invisible enemy and individual 360-degree fronts, a lot of people get hurt, but many unnecessarily. About three weeks ago we called in artillery air bursts on a bunch of civilians, killed three women and a child, and I talked in the dust-off chopper to med-evac ten more in more moderate state of mutilation. War is, just by nature, gristly, but this one is vulgar and dirty and guilty too. There's no justification for much of it... Let's just call it off and let the Vietnamese go back to their rice paddies...

★ ★ ★

    Game for another animal story? This one's about Pete the puppydog who turned up on the island where we unearthed the large rice caches; I think I told you about that. Pete's been with us ever since. He's about nine inches long overall, not nose to tail tip because his tail turns a full three-sixty over his back and it's so tightly curled that it would snap like a pretzel if you tried to uncurl it, just nine inches overall and about four inches at the shoulder. And awfully bow-legged. At first he was thin and very bow-legged, now he's fat and awfully bow-legged. His fur is like the nap of a paint roller for latex paint, sort of whiskery and about three-eighths of an inch long. His ears make a sort of half-assed flop when they're pricked and, when he's feeling soft and quiescent, his ears just stick out sideways, lending him an air of grave stupidity.
    About dogs in general in Vietnam; they're all of the same general cross-bred type, curly tail, bristly fur, and bent over triangular ears; they come in two colors; just off-cream and grey-brown (Pete's of the latter inclination) both solid colors, with no patches or gradation of tone. There are very few, or no, cats in Vietnam. I've seen one big tiger Tom since I've been here; he was probably exempt from death for food because of his size and evident toughness.
    Shortly after we got up on L.Z.Liz this last time, the lieutenant determined that Pete's breed was the famous Soona. The breed derives its name from its tendency to "soona shit in the bunker than outside".

That's Georgia humor; take it or leave it. Pete's young and his bowel control is not exemplary.

Pete has two garments, a double string of small beads for a collar and a sliced up olive drab sock for a sweater. A sock top was cut off, about nine inches of it. The ribbed-knitted top was rolled twice, turtleneck style, and makes an inch wide collar. From there back the sock is split down his belly for freedom. Just behind the collar, where the ribs give way to continuous knitting, there are two 'armholes' cut along side the belly slit and behind them, aft of Pete's ribcage, there's a bit of packstrap sewn to the sock to keep the sweater centered on his back. In his olive drab with his beads swinging, eating the best part of a lerp ration, he looks just like every other PFC in Vietnam.

This morning we were doing some rice paddy wading, about knee deep work, and Pete, of course, couldn't dog paddle the whole way. Some of our people slip him in one of the big, side of the thigh, bellows pleated pockets on their fatigue trousers when the going gets too rough for him. He keeps his forepaws and head out of the pocket, pushing up the flap to make himself something like a monk's cowl, while his head, bobbing and craning, takes in the view and the passing airs. My pouch pockets contain an air panel and a plastic bagged map of the AO, and furthermore the water was menacing them this morning, so I scooped Pete off a rice paddy dike where he was shaking himself dry, shivering (in the early, overcast mornings of the height of monsoon season, the air, not to mention the paddy water, can be quite chilly at five A.M.), and wondering what to do. "Surf's up", I said and carried him, his wriggling, warm and wet and bristly tummy in my hand, across a few paddies. Pup in one hand, radio handset in the other, panting a bit (me), squeaking a bit (Pete), my rifle hung horizontally just above waist level, rapping on my magazine bandoleer and my hand grenades. Then I passed Pete to J.R., hung my handset back on my packstrap, and took the weight of the rifle off my shoulders; I've added fifty feet of five-eighths nylon rope to my pack and any weight off my back and neck is appreciated.

I've received my package of goodies and razor blades, hat and beads. Everything has found a place (goodies inside me, hat on my head, blades in my toilet equipment bag with my razor and steel mirror), even the beads. You realize that every soldier has beads either around his neck or worked into the hatband on his bush hat -- every decent fad finds its way to the Army about three years after its peak. Just now aluminum peace symbol pendants, made in Japan, are all the rage with the machine gun set. Most of the beads worn are of local manufacture -- a string cost one hundred dong (one piaster) or two bars of soap and a pack of filter cigarettes out on the road -- and are of regular pattern and of relatively subdued color (the main principle of camouflage is symmetry of color and asymmetry of shape, afterall). I hadn't gotten any Vietnamese beads mostly because they were so popular and so meaningless -- the same reasons I use against a moustache or Volkswagen. Now I have some beads (tucked in my tee-shirt most of the time) with some meaning and I'm happy with them. Those who see them when they fall out of my shirt invariably say that they're distinctive and different and insist on a closer look.

The irregular sizes, shapes and textures of the beads suggests a wildly disordered, awfully unethical rosary (normally five Hail Mary's to one Our Father) demanding the utmost in prayerful creativity from the soldier fingering them in depths of two to three-fifteen A.M. watch.

That's the bead situation in Vietnam, thank you. I wear them secretly knowing that these beads mean something. In the dark watches of the night I also roll the deTomaso 'Mangusta' Corgi toy car that Rik sent me back and forth (very quietly) -- I sit squishing the suspension... up and down for minutes at a time, looking at it at eye level, digging its amber headlights -- but that's another form of devotion entirely.

Rik and David write all too regularly... both of them are obviously much more interested in my intellectual preservation than in the inactitude of my spinal column -- secretly so am I... They write all too often only because I write about once a month and then not very well. Mostly I'm afraid that they'll get bored with me being here and taper off this magnificent influx of stimulation. An M-16, handy as it is, is rotten company...

★ ★ ★

You really want to know about the M-16 rifle, fact and fancy?...

The M-16 in particular, is a new departure in weaponry, the largely aluminum construction and the all plastic stock contribute greatly to lightness at the cost of no other quality than longevity. No amount of preventive maintenance will reasonably extend the M-16's life expectancy because its demise is determined by the increase of tolerances on soft aluminum parts. Lubrication is good and the special lubricant supplied for the M-16 (LSA, a multi viscosity gel which looks for all the world like bacon grease) is in good supply and is liberally used. I've never seen a rifle which will keep shooting longer with reasonable accuracy on less maintenance than an M-16. No Springfield-designed rifle can match it for ease of maintenance, easy replacement of parts and handling ease. There the accolades stop.

The 5.56 cartridge is not a killing cartridge; it is a maiming cartridge. It takes three M-16 rounds to do what a well placed 7.62 mm bullet can do. Jungle/ambush war isn't the best place for well placed shots, nor does the basic training and infantry AIT make a good shot out of an average soldier. Even if a man is perfectible, it takes more than sixteen or seventeen weeks of sporadic training to do the job. So maybe smaller, lighter cartridges and a rifle which can operate as a fully automatic weapon when it's called upon to do so is the answer to mediocre shooters in a sudden and deceptive combat situation. The devil's choice.

But why a maiming cartridge? Here's some technical background: it's normal practise to keep the center of mass in a bullet as far forward as possible, to make it as resistant in flight to the effects of the wind as possible. In other words, the shape dynamics of a bullet and the raw effects of inertia are normally made as compatible as possible so that the bullet will tend to remain on course. The 5.56mm bullet in the cartridge made for the M-16 varies from that standard in that the center of mass is well back in the bullet -- the overall effect is that the bullet flies true until it meets a deflection at which time it tries to stay on course but will begin to topple or spin off its axis in flight. If, for example, in firing at a target the bullet were to hit a blade of tall grass, the shape of the bullet hole in the target would not be round, but would be a three-quarter view profile of the bullet. Making holes like that is called keyholing or sometimes, not to accurately, dum-dumming. You can read provisions concerning bullets in the Geneva Conventions. Last time I looked we were signatory to all the Geneva "rules of war" pacts up to 1956 -- but what's a hint of hypocrisy in a war with so many downright lies.

But what happens when an unstable bullet, such as the M-16 round bullet hits a target other than paper? Let's assume that the bullet has been true in flight and that the enemy wasn't in a stand of bamboo or elephant grass (which is highly improbable enough to be absurd) and lets assume the bullet hits a reasonable spot on the enemy, an arm, the collarbone, a protrusion of the pelvis. As soon as the bullet hits flesh it will begin to topple and spin, so a neat entrance hole is often accompanied by a very messy exit wound. A puncture in the front and something you can put your fist in in the back. But that's only fleshy tissue -- let's assume a bone in front of our bullet (our bullet? Well, yes, yours and mine, our...) the bullet in reaching the bone has to traverse some flesh so when it arrives it'll be toppling and spinning and, because it has been making a channel a good bit larger than its diameter, it will have given up a bit of it's energy (which because of the relatively small mass of a 5.56mm bullet (compared to one of 7.62mm) is not really much to start with) and instead of plowing through the bone it will be deflected off of or along the bone, splintering bone a bit (and often getting a joint) and giving the bullet a new avenue to travel towards its exit wound. A bullet in the upper arm often wipes out the elbow or shoulder, one on the collarbone (a mere nick) can put a spinal column out of alignment, and one in the pelvis (a grazing hit, again) can wreck a gastro-urinary system all to hell. Believe me...

 And now -- in the field with my twenty-three pound radio, fifty feet of nylon rope, five quarts of water, a couple of days worth of freeze-dried LRP rations, ammo, smoke grenades (4), frag grenades (3), trip flares (2), hard flares (1 cluster, 1 parachute), spare battery -- all on my back, I now have the largest traveling library in Vietnam. Not too sensibly, a helicopter dropped your parcel in to me on a re-supply haul. I'd gotten down to one volume of John Updike and a soggy tome called Politics and Literature (articles on Stendahl, Doestoevsky -- and so on), all in one pouch pocket -- now I have the contents of my rucksack hung all over my pack frame (socks on the radio aerial) and my rucksack full of books. Secretly I'd rather read than eat LRP's any day...

 Please bear with the contents of this letter. I get moody -- sometimes wildly optimistic, sometimes pensive (even reflective) sometimes...well, I do have local news and views to recount, I'll get around to them. Today we're having a typhoon and I'm holed up in a Buddhist temple, reading, writing, rained on... also, reamed, rifled, and wretching. (Or is that without a "w"?) God my head is slipping.

★ ★ ★

Today we're supposed to have a typhoon, one of the big blows of the monsoon season, so we're holed up in a decrepit and abandoned Buddhist temple. I'm writing in the very shadow of a yellow-robed and ceramic Buddha who's assumed the 'teaching' posture with his hands, but isn't really convinced that he can tell us very much. The alter is still hung with a few tatters of fringed silk hangings and some powdery dry flowers are crumbling in a vase on one side of the main image of the Buddha. There are a collection of imperial- robed, Vietnamese minor deity figures and a pair of auxiliary Buddhas, one in the 'earth witness' posture and one utterly reposed in an unconvincingly serpentine full-lotus, both of them with the traditionally depending ears and the flabby figure of the truly enlightened. A transistor radio is squawking distinctly un-ecclesiastically turned music from AFVN; tinny and undersized sounds somehow apropos. K.J. is heating water for a Lerp ration over little balls of C-4 (high explosive plastic, which burns with a sodium-yellow intensity) in a corner and L.M. is quoting over-loudly, in an atrocious accent, from a 1914 French grammar primer he discovered in a back room (the vestry?) and W.W. is reading old and wet letters. All fit activities for a temple, I suppose; enlightenment is what you make of it. This is the brand of enlightenment with mortared holes in the brick-tile roof and bulletted holes in the stucco-ed walls, rusty silk banners and the stink of incense unliberated so long it died of asphyxiation, rotting G.I. socks and a dissonant clang of machetes on dried bamboo -- to make tents of ponchos to meet the storm. Perhaps someone should draw the fine distinction between enlightenment and civilization for these people. "Hey Mister Tambourine Man" says the radio and the Lieutenant insists on joining in. Petey is roaming around with his sweater askew. If temples and typhoons, declaimed basic French and puppydogs don't sound much like raging war, just as well. My rifle, propped on its bipod, isn't really too attractive between the saggingly rotted, shutter doors of the temple nave -- have you desecrated a church lately? But the Buddha, I sense, would rather have me here in the storm. Of the soldiers, the Buddha needs me and I'm happy for the shelter and the temporal comforts of a rod and a staff...

★ ★ ★

...I've discovered why people with hot weather experience dislike chocolate: Hershey's Tropical Chocolate Bars. They are, I swear, sintered cocoa powder and portland cement, guaranteed not to melt in your hand, your mouth, your pocket, anywhere. Locally they're called John Wayne Bars -- they're that tough. They're also free, along with cigarettes and toothpaste and lifesavers, in the field...

★ ★ ★

    We're out here on this hilltop, CA'ed in about three days ago, socked in with weather, no re-supply and crawling going on all around us. The weather broke for about twenty minutes yesterday morning and in came a chopper (the most beautiful one I've ever seen) with water and a few lerp rations and the Chaplain aboard. All of a sudden it was rainy again -- we're up where the clouds sort of tumble into us and all of a sudden it's solid water and your cigarette goes out and paperbacks dissolve from both ends inward, either before you start them or before you finish. So I had a little warmth, fire going, nice hard wood like cherry and as convoluted as any coffee table bonsai tree, so over strolled the Chaplain, his spiffy fatigues slick wet, captains bars on one collar wing and a Christian Cross on the other, pasty faced, with a patent G.I. flat-top haircut, and spaniel-sad of eye. Mid-western of accent and very unsure of himself. Brigade had sent him out to spend a few days in the field with the troops. Such a chaplain, I said to myself, would be a blast to harass; I felt like sucking him into a commentary on Pauline interpretation,... and then demolishing him -- complete with footnotes from Deissman and the two verses from First Corinthians that I know. I was in a bad mood, let me tell you.

    Well, turns out that he was a non-denominational, do-it-yourself-interpreter, very much hip to the idea of individual experience and much too honest a guy to ever get stuck in the Army or a uniform with things on the collar in the first place. We took to each other like a sad-eyed not-finished-seminary chaplain and a haggard-eyed, eight day bearded infantryman, both wet and both thoroughly pissed off. We tried to put everything aside and just share the fact that we thought that Christian experience was attainable for different people in different ways, but that conventional liturgies were pretty shot as far as being a source of experience. We tried to share that and the fire and concentrate on that to the exclusion of the elements and our shoddier emotions. For a while we really succeeded. Then he started to tell me about the (I quote) "five rules set down in the New Testament, by Christ, through the disciples, for salvation, and I knew I was wet again. Felt like rapping my Deissman stuff to him again or whipping out my big, bold Episcopalian dogtags and waving them in front of his rule-ridden nose. We broke off quite amicably even after I'd enquired as to who held the rule book and the scorecard and what the penalties were for rule infractions? I just don't enjoy people telling me that anything is imposed on me in order to prove my faith in one way or another -- little children and irresponsibly deranged people and animals and, by God, even Buddhists, I think can have the experience of faith -- on a solely individual basis. So the Chaplain meant well when he started off, I think, but the necessity of having something to preach sort of did him in; the necessity of having to explain the experience forced him into a home-brew liturgy, but a liturgy nonetheless. He was a good guy, spiritually very well-endowed and physically impoverished.

    Now, about this having of experience: obviously you're not going the route of fasting or discipline to find an inner well of faith; you, quite rightly, I think, define Christianity as a social thing -- and since we're locked into a Christian cultural frame of mind -- it's in the social context that we seek THE experience. About group dynamics and retreats and group services I don't know much, because they're not much of an experience for me; either I feel the thing is artificial (the events of a week-end of slashing revelations seem rather meaningless when I discover my same old mind in my Monday morning desk drawer) or lacking in depth (where the sense of history is lost in the mumbo-jumbo of a liturgy) or poorly directed (in a canned experience like a very moving film). I'm a firm believer in Kurt Vonnegut 'karasses' of people and utter coincidence in experiences which are permanently stirring. I also believe (pretty much) that one's significant activities are not ones that are planned and well-or-poorly executed, but the spin-off productions -- an alignment of mood and talent and circumstances and time and place and available materials. Following that principle (or set of p's) can only lead to fragmentation in life in general. Following it just makes one circumspect and dillantantish and pretty useless. And sitting around waiting for an experience is, I'm sure you know, quite an unprofitable way to spend your time. But so is the conduct of life toward pre-planned (and, because they're thought through beforehand), anti-climactic experiences which just cannot mean much, but if sitting around, or just driving away tanks of gas (as I did as a younger seeker) in search of nothing in particular (which turns out is nothing at all) is unsuccessful and planning encounters and stirring events and setting up meaningful people is just as unsuccessful -- then, well hell, what does work?

 Well nothing 'works', because I haven't been able to extract five patent rules from the New Testament -- and I don't know what will work for anybody but me.

But since you press me (you do press me, don't you?) and since I'm in a benevolently preachy mood I'll tell you a few solutions of my own, for purposes of example showing -- when pressed as to what to do upon graduation Rik says, he tells me, "buy a motorcycle". It struck me that my response to people who give me the equivalent question in the Army, in terms of the Army is very much the same. I say, "Go to Jamaica and build automobiles", and when further pressed, I say, "Conduct a small business on the principles of living theatre and patent what I can". Note please that the words 'fulfillment', 'experience', and 'God' do not enter upon my expression of my life's uttermost goal. There's a secret there. Wanting to do something awfully badly and doing it is the most amazing experience. Ooops, a liturgy popped out. But I won't apologize, because it probably won't work for anyone else. Probably won't for me. Figure out all the things you want to be -- handsome as the 12-meter helmsman in a men's cologne ad, rich as Howard Hughes, excited as an operatic mezzo-soprano, dissipated as King Farouk, (God rest his soul), revered, healthy, artistic, woodsy, tweedy, preppy, polished, brusque, chic, knowledgeable, sought-after, loved, especially rich, hip, accepted, considered an eccentric, brave adored, regal, (a whole Roget's bunch of things you'd like to be); then know yourself well enough to throw out the reasonable and rational alternatives and shoot for the most absurd and ridiculous things and the rest will come naturally as will the magnificent onslaught of satisfaction when you achieve something totally alien to everything on the selected list and probably not even on the original roster. It's just doing something and being something. Just saying you know something upon which to base a personal choice... Plumbing the depths of self by projecting on a personal goal. By gar, what a pep talk! Now get out there and sell, sell, sell those Fuller Brushes, because you believe, believe, believe, in yourself (as the vessel for all experience).

There, there's my latest blow to the organized church. Tack that up on top of your ninety-eight theses, Martin Luther.

I'd like to spend next summer completely without obligation. Secretly I fear that liberation from the Army will only be a passage into a more constricting situation -- you must do something about (a) job, (b) school, (c) re-adjusting to life, (d) getting to Jamaica, (e) building automobiles, (f) etc. (g) etc. Terrible as I can make the prospect sound, I haven't considered re-enlistment yet. Rest assured...

★ ★ ★

    You must imagine a street gang ranging over their turf in search of the opposition, the opposition identifiable, the opposition wearing their colors, the opposition equipped (tire irons, cycle chains, etc.) and, thwarted, that same gang turns to whipping aerials off cars; breaking down grim, cast iron, decorative fencing; busting windows; slinging their chains around rattling lamp posts with all the might that they half-intended to use on the Reapers, or on the Comanches, if they ventured this far. Angry, and their hands stinging with the reverberations of their wrecking bars off the last hydrant, equally leery of having their patrol come up empty-handed and meeting a few too many Comanches, feeling the inexplicable combination of emotions involved in being out in force on ones own turf, being in the pink of condition because of a few beers, being more ready than ever, being cognizant of recent encroachment (an artfully primered Comanche Ford spotted on the block some little time before the beer), being spurred on by under-the-breath rock-and-roll all around one, being in full knowledge that this is it, if it will only happen. Keyed up, too much to explain, but the too-long tire iron now lighter than before, ever, and it can be held for a pool-table-perfect shot in the curl of the fingers, weightless and still commanding. Comanches got wheels. They're smart; just in and out. Like on our women. Whang. Another parking meter, another ivied arrowhead off a cold steel trellis around two o'clock in mercury vapor relief, grubby parking lot morning. Park all day -- five dollars

-- In the deepest, most unreachable central part of the night. Old people, smelling particularly old and alien, at bus stops and kids out of a movie, running real late because it's in this district, with bus fare and coffee money and worn-square lipstick, Kleenex and peppermint Lifesavers, not too choice possessions defining a life at a stage. Beating up on people. Locally we don't even call them village massacres. They're an expression of something you've got to get off your post-adolescent chest which is dirty and under-privileged through the offices of parentes who won't deign to be in loco. Ain't no difference between Detroit with tanks and My Lai with rifles, it's the same combination of ingredients and the same empty absence of a spontaneous spark of ignition, the together thrust of compression of expanding individuals stuck together by a defective but none the less binding adhesive. Shall we decide that people are brown or cream-colored or that they're April firsters or lucky-lucky October ninthers, or can there be more indiscriminate means of discrimination? IBM could just make some people haters and others hatees at purely random times for purely random intervals, like forever. We wouldn't even have to go beyond the premise that man gotta hate, individually, corporately, communally, nationally, racially, interstellarly, galactically, godly, eternally, and infinitely, then we could come up with a solution like war to take the problem to Duc Pho, instead of Detroit, Bu Prang instead of Berkeley, the moon, instead of Massachusetts.

    It's a phoney war when you go looking for it. Here in the mountains where I'm curled under a propped poncho watching the vertical sheets of wind propelled mist cloud the blue-willow-ware peaks, knowing it's rain from the patter on my soggy pack and watching the droplets form on the base cap of the frag grenade on my harness. Will we be mortared? Will the Comanches come on their bikes, in their cars -- wearing their colors -- or will it be another night of random firing while the clear rain of the mountains tries to rinse the rank muddy slickness and fecal tang out of darkly wet half-assed rubberized canvas? Don't read the Comanches as allegorical NVA, because secretly the Comanches don't exist; that way every grey '53 Ford can be a Comanche car and every old woman with a shopping bag can be a Viet Cong sympathizer. Come the rocket round and the Reapers will be real, come the NVA with bugles and drums and rifles and that furtive and unfamiliar face in the block's package store will truly be an enemy. It's a phoney war when you go looking for it. The war is in each of our perfectible selves, vastly more war than man has the munition-making capability to cope with. Astride a one-lunged snowmobile in the virginal tracts over which no person can take power, or on a peeling fire-escape landing when its 103 and every inch in view is owned, leased, sub-let, cooperative, brokered, Con-Edisoned, opted for, bid-on, cut-rate, pre-furnished, no down payment, in escrow, on the margin, and smells of spare-ribs, in the Carpathians, the Appenines, the Himalayas, here... There's the big potential for more war than you'd care to read about in learning to know your enemy and the big potential for more peace with more of resonant silence than the moment in the car when the out-put transformer in your four-track stereo tape player blows out (in the middle of the Warsaw Concerto). Even in these mere foothills I can look down on clouds, I can be in a cloud; I can grasp a cloud. And should we grab at moments of peace like that. Should we use the amazing capability of the human mind to selectively remember, to magnify, to add height and density of foliage to Christmas trees, to store and run together. To make peace which is as real and as tenuous as the cloud (see above) a mental reality.

All I can remember is the queasy feeling of walking into an unfamiliar and cheap restaurant and having all the eyes say that I'm a Reaper, or worse yet, a Comanche... And I remember the day, out on the island where we turned up a rice cache (21 tons -- a divisional record), when the little national policeman (your smiling gestapo) said that the line of moving people were VC ("Veecee, Veecee", he practically jumped up and down) and we called in artillery air bursts and long range machine-gunned them when they ran and I wound up med-evacing ten women and children and putting tags on the arms of four dead women. I guess we knew that they were more or less an evacuee party, trying to salvage rice from an adjacent ville; Jose, the arty forward observer knew, and Elrod with his jammed, un-jammed, jammed again machine-gun knew, and I knew, long before I called, "Dolphin-one-eight, this is Tango-six-two-India, I'm popping smoke for you." and the chopper came in after telling me "no more than two to a litter." The radio-telephone operator is custodian of the detainee and the dead tags. Four dead women. In Jose and Elrod and me there is the reason of peace and war and rioting and amphetamined-out-of-this-Godless-world hip people and militants and the conscientious objector in Bravo Company who refuses to carry a rifle and instead carries the mortar base plate, confusion and antagonism, and keeping it all bottled-up until it comes out like this in cliches or powdered eggs or Tom Wolf-ian sputters. To die, as graphically as an anatomy lesson centering on the articulation of the rib cage, from loss of blood. (That same night, three of our third platoon people were killed by a booby-trap while setting up for an ambush; one lived nearly a whole, precious, peaceful day, afterwards.)

★ ★ ★

...Christmas in Vietnam will prove to be quite something. We're scheduled to go directly back into the field, the base of the mountains, after stand down and my firm conviction is heavy enemy contact around, and on, Christmas; they probably won't even be able to ground a chopper with our individual dabs of boned turkey come re-supply time. It gets hairyer and hairyer all the time over here. And I don't mean just having to eat lerps -- I'm sitting here watching the traffic on a med-evac pad. I don't mean to scare; I'm spooked enough for several people; but merely that things are picking up in a mid-monsoon period...

★ ★ ★

...Huddled under my poncho trying to preserve the condition of my stationary, all thought of quality gone, writing away while monitoring my trusty two-way radio, looking out at the little plastic Christmas tree that one of our machine gunners received in the mail and planted before his draped poncho. Put the little metal car (a detomaso Mangusta) that I carry in my pocket beneath the plastic tree and, lo and behold, we'll have toys under the tree come tomorrow morning. All the amenities are not lost. One little tupperware container of mother's best cookies, too. No, all is certainly not lost at Christmas time...

Next Christmas eve I'll perhaps remember my rainy night, squatting beside my radio (on my plastic covered map to keep my bottom unsuccessfully dry), watching the bushes move, and every so often munching on mixed nuts (without peanuts!). Maybe this was the Christmas eve, and the Christmas, to make the rest worthwhile...

★ ★ ★

Dust the drifts off my Sprite and put a tuft of evergreen on its windshield for the season. Sprites are essentially Christmastime cars, being, as they are, giant embodiments of toys beneath the tree…

★ ★ ★

...I'm under a general cloud because everyone is down on me for not getting drunk like the rest of everybody and his assistant gunner. My narcotics and alcohol experience is ever widening but I prefer to participate only when in a celebratory mood and only in the company of friends. Can't drink for escape, though had four whiskey sours with the platoon sergeant to no avail; home brews, a plastic Colgate 100 bottle of Canadian Club and packages of powdered sour mix... Can't bear reality and the other world is no better. Lot of people smoke so much they have to speed... to see what the world looks like, seem quite normal all the time, nobody cares, everybody with a magazine in his rifle, safety catches on -- only bad when they're drunk and flip the selector from safe all the way to full-auto to enforce a point. Send me some daiquiris and I promise to make my impression only on semi-automatic or, at best, with a concussion grenade. Better massacre than Sand Creek... or My Lai 4..., all of Company D and all the tires of the Lambros in downtown Mo Duc and three water buffalo cows, six piglets and a forward air control plane. Headline in the Massena, New York Sneeze: PFC Bailey states he was frustrated -- banner heads, 158 point type, extra editions, kill the fatted calf, knuckle the heads of newsboys to steal extra copies to send to friends...

    Happy New Year -- Ah New Years, and kangaroo feet down Fleet Street and wealthy Percy's to give us all a new start, a rasher and well scrambled pair of eggs and wooly slippers these mornings. Sunny days on pinkened, Jamaica sands, sipping British West Indian Rum, selling Xerox short, only foggily 'membering our meager days, new hours and more fleeting and tasty moments of time. Rich and healthy and Christ, in Love in a new year.

★ ★ ★

...I've just been temporarily moved from my platoon leaders RTO slot to the company CP where I carry the battalion communications equipment for the commanding officer (...CP stands for 'command post'). So right now, New Years Eve, I'm sitting in the commo bunker on L.Z.Liz with the CO who's reading the 'end of the decade' issue of Time (Tokyo reprint edition) and every so often chuckling and reading scraps out to me. He so seldom permits a laugh that it's a bit of a revelation. The CO is a thinking sort of person who has an awareness of Vietnam nationally and culturally, as well as a clear vision of his mission, and everything is deadpan serious. If he were, at this time, to be separated from the Army, I think he'd rapidly find a place outside it where he'd be doing good (and win a measure of acceptance), remaining in the Army he's a real sanity force, fulfilling every bit of what's required of him in responsibility while sensing the peculiar balance of protection and pacification that the Vietnamese stand to gain through our presence here (in the ideal sense) and the terrible detriment that we stand to hand them (in the less exemplary instances of our behavior). A truly handsome man, smokes Pall Malls, is measured of speech, completely unflappable, a reasonable marksman (a clay pigeon shooter with a set of matched Winchesters), up to date on films, not condescending, tactful but not scheming -- I like the guy, accept that I cannot appreciate that he takes his satisfaction from assuming the command responsibilities of an infantry company -- he seems to be seeking a cold manhood while under-rating himself and underachieving. To each his own bag, of course, but I'm compelled to a value judgement analysis because I like the man and am troubled by the scheme of the Army. He's perhaps twenty-seven years old, or a touch more, un-married, though he'd make a positively brilliant father (my current image, at least), and very close to becoming a Major...

★ ★ ★

    I'm up on L.Z. Liz for New Years Day, enjoying the occasion and working temporarily in the Company CP, handling battalion communications. At CP level, commo has to be maintained both with internal elements on one frequency and with battalion agencies, air support, and re-supply agencies on other frequencies. In the CP we have a guy who carries a scrambler unit which plugs into my battalion radio -- daily we un-key the scrambler to a new pattern and have fairly secure, plain-talk communications. Up on Liz the duty consists of commo coordination in a radio shack, with both the radios and a set of land lines, phones, a miniature switchboard, and a cooler full of coke -- it's fun when everything gets going at once and efficient communications and exact relaying of messages conquers all...
    Perhaps you've wondered how the field soldier is re-supplied with razor blades and toothpaste and cigarettes -- apparently you have because you sent me a Dr. West's 'germfighter' and a tube of Colgate's ribbon. Well, every four or five days, with the ration re-supply, there comes a Sundries Packet, a plastic-bagged cardboard box containing three smaller bagged boxes, the whole thing maybe two by two by three feet, one per platoon. There are thirty razor blades, four pair of boot laces, three cans of foamy, four toothbrushes, six bic pens, four packets of envelopes, a couple of pads of unlined paper, bars of soap -- in another box is candy: the famous 'John Wayne Bars', Chuckles, Lifesavers, Charms, Chiclets, Kraft Caramels; and in the third box is cigarettes, some pipe tobacco, pipe cleaners, lighter

flints, book matches, and one packet of Beechnut chewing tobacco. So that's our sundry re-supply; about four packs of cigarettes, three packets of candy and generally whatever one needs in toiletries per man. Very neat...

★ ★ ★

    I find my issue... government sunglasses, though they have a good dark smoke tint, are too small in lens area; and glare from the hot, bright, tropical sun, which passes around the lenses, is very fatiguing. If a pair of Ray Bans .. is available with a prescription, perhaps you could... have a pair made up for me. I expect, also, that the flexible wire hook-type temples on such a pair of glasses would prevent the sweat slippage which makes plastic paddle-type temples so frustrating (like the time I stumbled and recovered in a knee-deep muck-bottomed paddy only to have my sunglasses not recover -- sometimes it may be tactically unfeasible to hold up the line of march while Bailey fishes in the mud and the crabs and the leeches for his glasses --) hot and hurting from the weight of the radio and the rucksack, helmet loose and trying to do a turtle over your eyes; Vietnam may be having a very informal and come-as-you-are war, but the little annoyances can get frustrating -er and frustrating-er and occasionally have grievous consequences...

★ ★ ★

    Ah, but the action is picking up; you'll recall my mention of a phoney war in which aggressions grow from the boredom and the muck and the tiredness -- now you may add to that the element of counterfire. The Tet Offensive, like the late, lamented Christmas season, has tended away from the holiday which inspired it -- this is the offensive, like silver street lamp garlands, going up a month in advance. The rationale expressed to me is that the opposition wants to overrun a forward fire base to have that success as a bargaining point for the Paris Talks (my, are they still going on...? with, maybe, third generation negotiators). More reasonably, one might suspect that the oppo. merely wants a piece of their geography back and four one-five-five howitzers as dividends. Or, they'd like to do a job on a company in the field to make life a little more comfortable for them. That's not a very strategic viewpoint, I know, but this is something less than a strategic war. I've retired my Aussie-style bush hat in favor of my steel helmet. There've been reasonably large attacks launched against L.Z. Charlie Brown, L.Z. Debbie, and Hill 285. They've been re-buffed. Our very own Bravo Company, strung out in a pacification action along QL1, has nightly contact above and beyond the average dose of snipers. I'm on L.Z. Liz right now, expecting, on the morning of the tenth that I'll be back in the field chasing Bravo's annoyance around...

    Please find enclosed department: here's a chieu hoi pass (pron. choo hoy) which is an item of air dropped propaganda which we drop into areas in which we imagine to be guerilla infested. Now, chieu hoy means "open arms", or at least the Vietnamese equivalent of that idiom. Any VC type person who decides to come across may do so by holding his weapon aloft and hollering, "chieu hoi", all the way. According to theory any U.S. military type may accept someone who chieu-hoi-es, and some of them do. But most chieu hoi's get gunned. Who likes a traitor?... Chieu Hoi's (types who've chieu hoi-ed) generally wind up in a liaison military outfit called, credibility gap, the Kit Carson scouts. They're attached to U.S. military groups, rather than ARVN's, as scouts, or if they have a knowledge of English, interpreters. And most chieu hoys stick with the KCS for a while, because they get an M-16, all the ammunition they want to blast off, and C-rations in the field. Kit Carson Scouts stick with the job a while and, when it begins to stick in their craws, they take rifle in hand and rejoin their VC outfits or the nearest VC operation. Some Kit Carson's I've encountered are adamantly anti-VC and really do scout out the opposition, do brutal interrogation (oh my, yes), and are swell executioners. They're a bit morbid, exhuming recent corpses (anything for a body count) and getting into the thick of the shooting. Our new Kit Carson is a kid named credibility gap Dung (somewhat silent 'd'), and George, our ARVN staff sergeant interpreter, says he's a good kid, real ex-VC (some chieu-hoi's are, of course, merely starving, or fed-up, peasants of a civilian or essentially neutral (if the word still has meaning) character) and a good scout. His rifle is as big as he is and he looks about thirteen (which is the abrubt end of Vietnamese adolescence). I expect that he'll soon be using his new rifle from the other side of the hedgerow once the offensive gains momentum and a personality. There's an allegory there about the Vietnam War -- just swap sides awhile to take a breather; the Americans just keep crawling in the muck, ever onward, ever onward, to where we've been before (and lo, and behold, even pacified) and all the while the guerilla people are taking things easy to pace themselves for another twenty -five years, or a hundred years. They've been at war since 1291 A.D. anyway, what's a century, or a millennium. Ah, oriental measured culture. Perhaps we'll depopulate the NVA and be able to call it victory when they no longer field an army, but by that time they'll have 'defeated' the U.S. military by having them called home on account of un-rest at home (picture yourself a military man in a war with 'God on your side' to fully appreciate that), a falling from grace, and total disenchantment on the public part. The North Vietnamese claim victory, the American military claims victory within its grasp, (we'll never know, because Mongolia is a big place to depopulate), and the guerillas, who've been vacationing off and on the whole while, merely go on about their business, brigandry or patriotism, who knows, who cares.

So raise high your chieu hoi pass and, holding aloft your .22 rifle (which I trust is properly registered), turn yourself in to the nearest cop and take a well deserved rest (as I am), we have a lot of work before us.

★ ★ ★

    Duc Pho they tell me, is Ho Chi Minh's birthplace. It's also under the name of L.Z. Bronco, our battalion and brigade headquarters. Can you verify the place of Uncle Ho's birth? I'd be interested, because Duc Pho would be a hell of a place to have a last ditch fight... I'm but a handful of kilometers inland and a bit north of there. Not really in the central Highlands...

★ ★ ★

Once again hiding from the elements under a poncho, just finished reading Faulkner's The Reivers (it's taken me ten days, off and on); I'll write in reply to your letter of the thirteenth. I do have other letters to base replies on, rest assured, but I have to be in different moods for different people (after months of no telephone to rely on, letter writing becomes a science, nay, a discipline, an art of overcoming lapses in a talent which occur according to no geometric pattern). Reading Faulkner leads one to over-use of the parenthetical.

Yesterday -- yesterday was a slow day up to the last minute. My platoon, which has been operating separately from the rest of the company for a week, sat around most of the day waiting for a CA... All day, sitting reading, eating C-rations in the rain in a hastily prepared perimeter around an equally hasty landing zone. Any minute now -- for hour after hour. At five o'clock it was decided, at higher levels of command, to give us our re-supply at our pick-up rather than at our drop zone, because re-supply is strictly a daylight activity, whereas CA's can go on until the last throes of twilight (or into the night). Our re-supply helicopter came in bulging with cases of C-ration, rubber water jugs, a sack full of mail, and some drop boxes of hot food. Lots of extra re-supply material because it was not expected that it would be possible to give us daily, or even every other day, re-supply in the position to which we were to be CA'ed. We each grabbed a whole half case of C-ration, a couple of cans of cold ginger ale, a bite of hot food. I received a backlog of two letters... and a printed tee-shirt from Kieth Black Racing Engines ('Black Magic') which I had sent away for in November. Then, with this heap of un-opened rations, cans, and mail beside my rucksack, the radio brought word that helicopters were at last free to move us out on our CA. Breaking open C-ration boxes with one hand and dumping the cans into the top of my rucksack, I called in a helicopter to provide backhaul for our emptied re-supply material. Most of the hot food was dumped, a match put to the drop boxes, the water jugs emptied, and the rubber and nylon bags of ammunition and grenades and still iced soda dumped or prepared to be, in part, at least, returned to the support pad. The backhaul helicopter came in, accepted our empty jugs, bags, and an ammunition bearer going to the rear for a promotion board, and it left us still cramming our rucksacks full. Right behind the back haul helicopter came the first two CA helicopters and, still stuffing cans and mail into my pouch pockets, trying to hold my radio aerial down and my rifle up (out of the swing of the main rotor and out of the wet, whipped grass, respectively), I ran to the first helicopter and sort of threw my rucksack through the side door; since the rucksack was on my back I followed it with a half twist and wound up sitting on one leg (the other still outside, foot on the landing skid) as the helicopter scooted away. 'Scooting' is the only term I can apply to flying a few feet off the ground at only thirty knots (which never fails to amaze me). We are served by little UH-1-H 'Iroquois' helicopters (commonly called 'Hueys' or 'Slicks') which have a payload of six infantrymen in full gear in addition to a crew of two pilots and two door-gunners. Rather unnecessarily one of the gunners held onto my rucksack, perhaps to re-assure me;

but some mysterious force, like the one which keeps motorcycles balanced, prevents people from falling out of helicopters. Needless to say, the helicopter gained altitude (even under the weight of canned goods) and dropped us off at our drop zone. I plunked out of the hovering helicopter, kept my knees from buckling, ran, actually ran, about twelve feet from my plunk point before the down draft (contributing the final thrust to my weight and momentum) pitched me through a full somersault which shucked off my rucksack and gave me a black eye when my helmet flipped over my head. I came up, recovered my rifle, felt around for my glasses (found them) and became part of a defensive perimeter for the further 'lifts' of incoming people.

It was then, while ministering to my eye, that I read my mail and checked over my new tee-shirt (with which Keith Black had thoughtfully enclosed two stickers bearing his name and logo -- one of them will replace the Bell Helicopter Co. sticker presently on my rifle stock because I'm rather down on helicopters right now, at least until my eye unswells.)

To draw yesterday to a close: we walked about two kilometers to an ambush sight (with all the groceries and in near total darkness through knee-deep paddies). Around eleven we sprang an ambush which cost the opposition about seven hundred pounds of rice and one partially empty rifle magazine (which they elected to abandon for freedom of escape), then moved our ambush (for fear of repercussions from our earlier betrayal of our position) and spent the rest of the night quietly. We moved out before daylight and now, sitting around again, depleting the quantity of cans to be carried, today wears on toward tonight's ambush...

★ ★ ★

...Note: one 'piastre' (equivalent to $.85 to $.90) is divided into a hundred 'dong'. On the strength of that fact and the fine scenario I've read on Bill Calley's ("the second lieutenant with no name") first film, "The Good, the Bad, and the Mildly Paranoid", I hereby authorize a pre-production check on his slightly risque sequel "A Fistful of Dong" and the promising "For a few Piastres More". I've hired one hundred and two hungry Vietnamese and I think we can borrow Sam Peckinpah's effects man who did such fine work on "The Wild Bunch". I've also found a persuadable supply sergeant who offers me cut-rate black market prices on a case of cal. 5.56mm tracer ammunition and a box of fragmentation grenades. You liked "Bonnie & Clyde"? "Hang 'Em High"? You'll love to try a little tenderness, Bill Calley style!

Seriously, there isn't going to be much in the way of books coming out of this war -- the instant media of television precludes any conventional treatment, to include Joseph Heller pathos jobs. But there's a great film in this war -- something which manipulates Leesville, Louisiana against Mo Duc, RVN. Now that would be a great undertaking for a fledgling automobile manufacturer...

★ ★ ★

...Action is picking up in I corps and, against what I'd consider to be better judgement, the opposition does seem to be planning on something solid for Tet. More booby traps and such in evidence now, more semi-clandestine night-time re-supply activity (every so often, with increasing frequency, some bumbling VC's get ambushed between Mo Duc and the foothills and are found to be carrying new clothing, AK -47 magazines in abundance, raw explosive, and, of course, rice). We're doing a land office business in dust-offs and captured weapons...

★ ★ ★

...Now that the total cloud cover of the monsoon is giving occasional glimpses of sky at night; once again it's clearing and getting hot; the contrails of flights of high level, multi-engined bombers creep by the moon, westward to Laos and Cambodia, then eastward again before morning. The callsign for the B-52's is 'arc light' and I await the opportunity to direct an arc light mission which has been given a 'divert order' from its primary target to a 'target of opportunity'. More kilotons of conventional explosive power than was experienced by Nagasake -- right here on L.Z. San Juan Hill, L.Z. Bronco, or, if I wanna do my bit for military-civilian relations, right on scenic downtown Duc Pho. Oh wow...

My dear Rik —

Sorry to have been so tardy in writing. The monsoon in Viet Nam is tapering off, giving an early promise of spring for the northern hemisphere. But, of course, with the end of monsoon comes the rising activity toward a Tet offensive, and the action has been noticeably accelerated locally. That's my excuse; that is, to be taken for what it's worth.

You write that the Fiat is clunking along, but more recently Maris has written that it's died some nasty death and that you're now looking at, I quote, "a 1964 Rover." She notes some reservations on your part. Now, unless she means a Land Rover, I suspect that you're looking at one of the "3 litre" Rovers instead of the newer "2000" models. I've always thought that the 3 litre had a great deal of "car-ness" to its appearance, externally, and very sumptuous interior appointments, and an engine with a rock-solid bottom end (that is to say: bearings). Pending condition and mileage, I'd recommend one of the 3 litre cars. With the letter time-lapse I have no doubt but that my recommendation falls on the ears of a new crew-cab Volkswagen pick-up (with conestoga canvas top) owner. If you can get it started and can bear the cold you may run the Sprite, remembering that it's wholly uninsured, until such time as you can see fit to get a car or a Gold Star-model BSA.

I trust that your new friend, Susan, can adapt to the pillion seat of an over-500c.c. bike?

I'm very worried that the bugeye is going to ruin (or, rather, to worse ruin), mothballed under a snowbank. Do give it an occasional look (antifreeze, oil, and a shot of Christy Dry Gas) and, if you have the means (another car and cables) at hand try to boost it into life and run it until the temp needle is off the bottom peg. If worse comes to worse and nothing avails come warmer weather, we'll present the Sprite to Pandolfe at fits for a proper burial in June. But I would like to have it to use while Porsche-hunting — and a whole remanufactured Sprite engine is just $164.50. Stebbins would surely paint it for under a hundred dollars, and I have that megaphone resonator still to hang on it. In any event, I'll have my memories of its fantastic reliability, unblemished, and it's amazing personality, unvarnished.

I understand that I have Paul to thank for putting the snow tread tires on Maris's car, and for a variety of other appreciated actions for M. Give him my regards, please. I'm glad that your association with Paul has gone on.

I think that you've noticed that final exams and love and war put a distinct crimp into one's creativity. Any feelings other than frankly dispassionate toward subject, audience, or emotional bearing tend to leave one unable to write or, when pressed, to make the most horrible mess of things. I find myself able only to write "Viet Nam situation" monographs and analyses for my friend David, for whom I harbor respect and amicability, but no more. For my correspondents, especially for Maris, I find it nigh onto impossible to dredge up pleasant fictions, interest in the latest in Israel or from Miranda, the frontiers of General Dynamics, Connecticut General or City Cab, or beautify or expand upon the marvellous job done by television in portraying people being rapidly converted to hamburger. Hemingway's prose and Life photographs can tell you as much as I can now about war. Maybe sometime in the future I can catch some more sensation some new emotional way, in some other media. But for now I must say that blood looks like blood on a cut finger and one learns to hold ones stomach in the face of gushes of it much the same way one learns to put up with the reek of cat droppings in ones bathroom. Real war is just like war in any of the communications media, only, of course, it's real -- and that reality, like being plunged completely into ice water, is the difference between warmly regarding a frozen lact lake and being submerged in it, warm quilting suddenly soggily frigid. But reality is not the stuff of communications; accomplishing ones ends while skirting the real is the whole thing, from the smallest figure of speech to the most elaborate stage set. So, with only a measure of reality, or call it credibility or involvement or personal identification, to put across, what can I make of my communications? What could I say in a letter that would do the job of my looking at Miranda's kittens and digging on them, if indeed it turned out that I did on little cats. I expect that you know what I'd do in the ~~presents~~ presence of your cats -- and I can't make any greater ~~think~~ ~~thinks~~ thing of my misunderstanding appreciation of them by remarking anything other than my congratulations herein. Likewise, how am I to convey that to you, who are still utterly wrapped up in the aftermath of the kitten happening, in words of honesty and inoffensiveness, as I would react to you were I present at the momentous occasion and through the ongoing existance of many, many kittens? Further, how am I to overcome the burden of daily dustoffs and captures and booby traps and what-not and make myself back into the person who would be present to dig on your cats? Or on your new friend Susan? Or doings at the old Corgi shop?

I've got this impossibility of situation, audience and emotional state to get around before I write, and, as is reasonably obvious, I can't really get around it. As one who's dealt in the devious realms of making people suspend their disbelief by artifice, I'm sure that you know that the old devices break down before something or someone or some feeling which demands to be dealt with in the honesty of sincere and strong feeling. I'm willing to do many a put-on, and create many a fiction, and sham a great deal of cocktail party interest -- for people and situations which can at best amuse me; but, Christ, Rik, I care to do something better than a credible acting job or deceiving job or feigning of interest for you and for Maris and anyone who I <u>know</u> rather than just <u>perform for</u>.

It's bitterly hard for me to think that I cannot project something of an understanding better than a tale I know you'll see me, in, across eleven thousand miles and a few months. But I see myself slipping a bit in your conceptive field -- and I hate to see myself, bad or good, growing fuzzy around the edges.

Especially when it's me who's looking. Bear with me man, this isn't a plea; it's something of a statement of my communicative limitations, re-evaluated. Something touching on absolutes. I'll be there, sitting on the cracked red leatherette, strapped-in, and rowing along with the gearchange lever, but until I get there I won't be there. You know me in terms of reaching to the rusty little car, but that's not me in gravy-like paddies, squashing along, or me coming out of the saw grass, iodining my arms and face. Maybe, if I could render the context, you could get me. I'd almost rather you wouldn't grope for that context.

Anyway, my little mangusta, still in my pocket, now lacks an appreciable amount of paint and, after the current of the Sông Trà Câu caught me and my rucksack, it's had an unrinsable bit of mud in the cockpit for a while now. But I care for it, as I'm <u>sure</u> you know.

Surely,
Rino.

in the rocket pocket
February 2nd, 1970

★ ★ ★

    I have recently been cataloging, Maris, all the things I haven't done recently (among them writing letters, but I mean on a grander scale) for example: I haven't eaten with a metal fork since the twenty-eighth of September, nor flushed a toilet since the sixth of October. Someone had an odd dogtag around Pete's neck and I bent to read it and the dog craned his neck so that I'd ruffle him behind the ears -- it was such a familiar thing to do, to scratch a dog that way and yet so peculiar, for in not having done it in months, I'd forgotten the sensation and response. That set me off -- I haven't had any but reconstituted milk, nor a bottled coke, or oral discourse with an intelligent or even interested person; been clean, nor been unafraid, worn undershorts -- I just took my boots off for the first time in a week (and given a good spot of time on Liz I'll get rid of this ringworm).

★ ★ ★

...On Vietnamization: you have seen the flaw in the system -- they just don't care, only we care to maintain a stand-off or continue to flush guerillas and maintain 'pacification' forces. It's our war, not theirs, and it's their country, not ours. Very strange, very stupid at times when someone is hurt.

★ ★ ★

    Well, my dear Maris, here it is St. Valentines Day and I'm once again out in the field. After a week of dribble and drizzle it's pleasantly sunny today and I'm working on my tan. After a couple of days back in the field my kidneys are again used to my rucksack frame and C-rations are beginning to get boring. Fortunately I received a parcel from Mother today, so I can fill-in my nutritive gap with S.S. Pierces canned strawberries, shrimp and an Edam cheese. Ah civilization, thy name is date-nut bread.
    Let me tell you about micro-frogs; it's not much of a story, but now is the season for the little frogs to come out of the paddies and the little things hop around on a very small scale do to their small size -- a half inch, legs not extended, they look like a bit of Tiffany's best enamel work scatter pins. One of them always hops on your face at night and then, scared by the unexpected texture, it takes two minuscule hops to get down. I try not to start when one plops down, but it's difficult to get a little frog belly in the ear and remain completely calm. Makes one feel at home in nature if one can relax and let the frogs hop, though. Also mice and mongeese. A mouse tried to bed down in my shirt the other night, cozying up to my navel, but, scared of squishing him, I took him out and offered him my towel, the one I drape around my neck to cushion my pack straps. I don't know whether he stayed; very bold these wild mice. Mongeese are like big rats or opossums and not very pleasant or excitingly different, but I thought you'd like to know I checked them out. Also saw an enormous

python the other day -- those exciting animal names grow a bit meaningless or prosaic when we think of them as automobile models or can opener trademarks -- but a real python opens your eyes and gives new meaning and respect to the name.

Thinking of micro-frogs and other hopping beasties -- I guess a cricket or grasshopper with its exoskeletal and disproportionate legs is more efficient a jumper than a frog, more bounce to the unit of energy expended, but frogs are so much more organic in their styling. Like some older cars with hips and bulges instead of bowed planes and choppy scoops like newer ones. No doubt the squarish ones, like crickets, move more efficiently, but give me a well-faired in frog or a coke bottle car any day. Viva frogs. And bottled cokes, too. Golly I miss them. Walt Kelly does good frogs in Pogo -- my, I've been thinking a lot about frogs lately.

...I have a new fantasy -- I pretend that I'm a Belgian mercenary and this isn't my war, I just work here. Lieutenant N. is the taciturn Pole who stuck it out to the end in Biafra and SSgt. W. is the disinherited German junker/warrior who pulled out on Gamel Nasser when he saw Israel winning. We have fabulous games. But Captain W. still doesn't know that he's ex-Major W. who lost his command in the Congo. We're all reprobates and lost causes of one sort or another. And we could use a Swedish relief mission, I assure you.

★ ★ ★

...Now to the matter of 'Marvin the Starvin' ARVN (remember, twenty years ago, the ROK's (Republic of Korea), well, now it's ARVN: Army of the Republic of Vietnam, pronounced 'arvin') and, can the Vietnamese Army take over the military responsibility here. Well, no, they can't -- they don't even want to. The scheme of Vietnamization says, "give the ARVN's the helicopters, jet fighter bombers, close air support, and the big guns and, with all our advantages, they'll at the very least, be able to maintain the stand-off situation that we've achieved. That's the theory anyway. Right now the ARVN's locally don't have any artillery; they rely on our helicopters and pilots for helicopter assaults and for what aerial re-supply they get; occasionally one of the old washing machine Skyraiders that the ARVN's have chugs across the sky, very occasionally, mostly our Army and Marine Corps Phantom jets work out their light bombs and mini-gun pods; the ARVN's have no gunships (helicopter Cobras and Sharks), no med-evacs, no flareships, no Cargo Chinooks, no trucks even. Of course, you say, neither does the VC or NVA have those facilities -- very true. But Marvin the ARVN, who bums his food and ammunition from you, because ARVN supply channels are pretty corrupt, has come to rely on jet planes and C-rations and big guns (laid-in by good cannoneers) when he calls for them on his American made radio behind his brand new M-16. Marvin is in no mood, at this late date, to lower himself to fight, quick and dirty and hungry and at night, like his more patriotic and dedicated opposition.

The Viet Cong guerilla, or the 'uniformed' NVA soldier (last outfitted three years ago) is fighting against that iron-clad might (and doing decently still, thank you) and the corruption of his brethren who are sitting idly by and watching the strength of their own nation be sapped by indolence in the face of free war machinery that cannot be bought with love or money and is growing increasingly less-freely given (as the U.S. Army pull-out gets real -- broad hints, no real action, yet). The ARVN's don't care enough and have been so hurt by assistance that, when the Americans go (if they ever do) Vietnam will not remain North and South, but will be unified under Hanoi and the NVA. That's sad, but Southeast Asia is doomed and we cannot give the might without hurting as much as we temporarily help. It was a bad game from the start.

★ ★ ★

...The Plain of Jars seems quiet lately, in the papers, that is. The B-52's still go over every night. Quote: "We have one-thousand forty government employees in Laos, no ground troops among them, and no ground troops committed." But how does one go about holding ground gained by strategic bombing? Wonder how Laos looks this time of the year?

What price glory, now,
Captain Flagg.

★ ★ ★

    I have to break it to you; I lost my trusty Swiss Army knife some time ago and I've been relying on a little buck knife (looks like a filleting knife) sent me by my parents. The corks I've pulled and the doors I've latched with that now lost knife make me very sorry I lost it. Possession is a tenuous thing in Vietnam; nothing you can depend on, really, just use what you've got while you've got it, plan on losing it in the dark and damp. But my Swiss Army knife I'd rather not have lost.
    ...It's been an experience to have gone twenty-one months without a dictionary. I have to guess at information and I can't find out what 'austentitic' means or the difference between a plain bearing and a bushing. Above all other material things, and I'm so oriented to them, that it's a distinct loss, I miss the books. Is it a 'Florence' flask? Or a 'Milan' flask? Some Italian town.

★ ★ ★

...Things have been very much a matter of mood today and lately; ...I feel quite out of touch or something and, not thinking to very much of my past twenty-one months' experience, especially the Vietnam episodes, I'm left with none to vivid or to vibrant two-year old standards to apply. Entering the Army I wondered how this would feel when the moment arrived -- egad it's bitter. I wondered if one could retain a continuum of life despite the hiatus of some processes. Apparently not and I'm to join the savage souls, newly outfitted in latest styles, new job and troubled with the vocabulary, who roam about saying "You were at Tay Ninh? I was Amcal at Duc Pho", or something like that to the most improbable people.

I'm writing by the light of a 'dink' candle stuck atop an eroded human (human, I emphasize) skull in my bunker. Not a skull as the philosophers kept in their warrens to remind them of mortality (although it surely does do that), but just a human skull with a fat red candle dribbling on it, common to us all, just more antiqued; Jerrie carried it in. Man, it just has to change you. Empty Coke cans, fragmentation grenades, three bars of institutional Lifeboy, L-head flashlight, a month old Stars and Stripes on the table, too. Big beams and sandbags in flickering light. Burn-scabbed shelf where I sleep, wet poncho liner and soggy Road & Track. A humble existence and not a spiritually elevating one, despite the word of Baba Ram Da, ne' Alpert, or any of the lamas, bodhisatvas or dervishes, ones in common context I mean. Quite ungodly, quite effective.

"... the glass is falling hour by hour
The glass will fall forever.
But if you break the bloody glass,
You won't hold up the weather.'"

    Louis McNiece... "Bagpipe Music"

★ ★ ★

    Ah, what delicious morbidity I entertain this night and soon I'll stand guard on top of the bunker.
    The Mangusta? The sand has at last been rinsed free; the metal, variously bereft of paint is attaining a polish; and the window glass, ground to opacity, is neatly holed in one rear window, a scale-sized break. Some might say it gaining in character, others might think it quite doggy, depending on how long one has left in country.

★ ★ ★

...Our last company commander, J.P.W., Captain, Infantry, was quite a decent sort of person. Though soldiers are a bit hard to respect, it's an easier matter to respect a professional, especially when he acknowledges his mission to be objectionable and absurd and views Vietnamization as a grim joke, and sincerely, but with professional quietude and efficiency, attempts to alter things. First successful man I've seen in working from the inside out on the problems of Vietnam, in the Army or anywhere. Anyway, I spent some time with him while temporarily assigned to the company CP on the battalion radio.
   ...As much as a PFC and a Captain could get along, we got along. He was injured a couple of weeks ago and I'd felt as if I'd lost a friend. Up in Chu Lai in December I'd bought a copy of, ah, the Ginger Man to be re-read and, at my offering, Captain W. borrowed it, quite enjoyed it. Everyone else in the CP, knowing him only as a stone face, was quite amazed with his amusement; only I already knew that he laughed. With Patrick for a middle name? Now, after a few months, my other ranking friend, Lieutenant N. (Providence College, 1968, home in West Haven), our artillery forward observer, whom I'd already lent The Saddest Summer..., is working on Ginger Man -- so Donleavy comes to L.Z.Liz in a big way; I have great hopes for N., despite a Jesuitical or Benedictine or some sort of education, he has the proper sort of insanity. We used to work on captain W.'s mind by staying up 'til all hours (taking everyone's radio

watch) in the radio shack arguing man's fate (he's an evolutionist and I'm sort of a gradual revolutionist) while having beer drinking and gum chewing contests (my record is twenty-four sticks of Juicy Fruit and three chiclets (peppermint) while N. has gone twenty-eight sticks of Teaberry at which time he fainted due to jaw fatigue and nine concurrent cans of Falstaff). All the while, though, we were amazingly lucid.

★ ★ ★

    The company left Liz on the twelfth and since then we've been working an area south-east of Bronco (Duc Pho), called the 'rocket pocket', because it's the place from where the rockets come when Bronco is rocketed. Walked out on a patrol this morning and, after getting some little distance out of the perimeter we heard a boom and then radio traffic to the effect that another patrol had hit a boobytrap and requested med-evac for three people. As it worked out they only needed a dust-off for one man; the other two where KIA. In not a little trepidation, my patrol has held up and we're just sitting this one out...

    Your good humored letters... are very uplifting for they speak of the restorative power of rawest freedom. When I'm down, and I am so all to often, I blame my down-ness on a great variety of things, mostly internal ones, and seeing you, suffering the same sort of separation and fragmentation I am, still quite creative and in decent spirits gives me great faith in the powers of self determination over the blind staggers of mind, as incurred while in government turpitude, er, service...

    I must stand guard now. Adieu, amies ... and bear in mind that today I am a 'two-digit-midget' with but 99 days to go. Did I tell you my R&R runs from the thirtieth of March to April fifth? It seems attainable. So must all of life.

★ ★ ★

...You must realize that these letters are written one shot and often without more than a half-assed spelling check. Generally they're written in rain or among noisy rats, by the light of two candles, in a bunker. Save your tears. No apology, but realize that I fantasize enormously, that I insult, or less intentionally offend, bore, infuriate, and, like this, lose my cool, quite without more reason than an intention to fill pages with something that will keep me alive in minds. To be remembered is an honor and the whole of my object. I should exult, but I shudder to think of my moods and mind changing, even in my conservative inertia, and the no less than comic character my letters must cut me as.

The deTomaso Mangusta

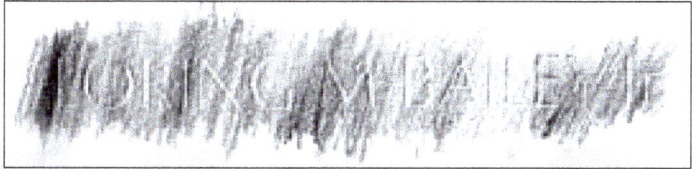

www.ingramcontent.com/pod-product-compliance
Lightning Source LLC
Chambersburg PA
CBHW040324300426
44112CB00021B/2870